OBSERVATIONAL NEUTRINO ASTRONOMY

*Proceedings of the Workshop on
Extra Solar Neutrino Astronomy*

OBSERVATIONAL NEUTRINO ASTRONOMY

University of California, Los Angeles
30 Sept – 2 Oct 1987

Editor
David Cline
University of California, Los Angeles

Published by

World Scientific Publishing Co. Pte. Ltd.
P O Box 128, Farrer Road, Singapore 9128

USA office: World Scientific Publishing Co., Inc.
687 Hartwell Street, Teaneck, NJ 07666, USA

UK office: World Scientific Publishing Co. Pte. Ltd.
73 Lynton Mead, Totteridge, London N20 8DH, England

Library of Congress Cataloging-in-Publication Data

Observational neutrino astronomy.

1. Neutrino astrophysics – Techniques – Congresses.   2. Radio telescope – Congresses.   3. Supernova 1978A – Observations – Congresses.   I. Cline, D. (David), 1933 –
QB464.2.024   1988   522'.686   88-37831
ISBN 9971-50-823-0

Copyright © 1988 by World Scientific Publishing Co. Pte. Ltd.

All rights reserved. This book, or parts thereof, may not be reproduced in any form or by any means, electronic or mechanical, including photocopying, recording or any information storage and retrieval system now known or to be invented, without written permission from the Publisher.

Printed in Singapore by JBW Printers & Binders Pte. Ltd.

Introduction to
Observational Neutrino Astronomy

The observation of the anti neutrino burst from SN 1987A can be considered as the start of observational extra solar and extra galactic neutrino astronomy. It is hoped that neutrino signals from astrophysical sources will be detected in the future. The meeting held at UCLA in October 1987 was organized to discuss the prospects for this to happen and detectors (telescopes) needed to observe such sources. These detectors fall into three categories that are related to the energy and expected flux of the neutrino or anti neutrino sources

(1) MeV neutrinos or anti neutrinos from stellar collapse requiring very massive detectors with low energy threshold. The Kamiokande and IMB detectors are extent versions of such detectors, but are too small to observe stellar collapse beyond the local galaxy and nearest galaxy (LMC). The LUD and ICARUS detectors at the Gran Sasso are a new type of detector that will be sensitive to galactic SN.

(2) GeV neutrinos from various processes including SUSY cold dark matter annihilation in the Sun. These detectors require very large areas and modest threshold levels. Detectors such as MACRO in the Gran Sasso, and a new generation of water detectors (such as a High Lake detector or the LENA detector) with very large areas ($> 10^4 m^2$) are examples of such detectors.

(3) TeV neutrinos from astrophysical sources due to particle acceleration processes in the large magnetic fields. The Lake Backal, DUMAND, and GRANDE are very large detectors that may serve this purpose.

Ultra high energy neutrinos have been searched for by the Flys Eye detector in Utah, with null results so far. At the meeting we heard of several calculations for the neutrino flux from these various sources, and the latest news concerning the neutrino burst from SN 1987A. There was a discussion of the limits on neutrino fluxes from SUSY particle annihilation inside the Sun, as well as exotic sources such as Cygnus X-3. There was also a discussion of the current limits on the electron neutrino mass from detection of the anti neutrinos from SN 1987A.

The major conclusion of the workshop is that a new generation of neutrino telescopes is needed to enter into this interesting field. While several candidate telescopes have been proposed or discussed (DUMAND, GRANDE, LENA, LHD, etc.) it is not clear that the levels of funding required are going to be available.

David B. Cline
UCLA

TABLE OF CONTENTS

Introductory Remarks

    Introduction to Observational Neutrino Astronomy
        D. Cline   v

A. Neutrinos and Gamma Rays

    Supernova Neutrinos
        Adam Burrows   3

    The Nineteen Neutrinos of SN 1987a: Mass or No Mass?
        Terry P. Walker   9

    Cold Dark Matter: A Source of Energetic Particles
        Joseph Silk   15

    Cosmic-Ray Neutrinos and Other Stuff (Antiprotons, Gamma Rays and Positrons) from Galactic Dark Matter Annihilation
        F.W. Stecker   23

B. Detectors Under Development

    Neutrino Pulse from Supernova Inner Core: General Comments on the Microphysics and Its Implications for Pulse Duration and Spectrum
        R.F. Sawyer   59

    DUMAND and Neutrino Astronomy
        V.J. Stenger   76

    The Sudbury Neutrino Observatory
        P. Doe   92

    Neutrino Detection with MACRO at Gran Sasso
        Charles W. Peck   106

    Characteristics of the "SMART" 35cm Diameter Photomultiplier
        D. Samm   109

C. Results from Existing Detectors

    The Mont Blanc Detection of Neutrinos from SN 1987a
        P. Galeotti et al.   123

Search for Neutrino Sources with the
FREJUS Detector
H. Meyer — 134

On the Possibility of Detecting Solar and
Supernovae Neutrinos with IN^{115} Detector
A.K. Drukier — 147

Comments on the Neutrino Bursts from SN 1987A[1,2]
Arnon Dar — 164

D. Theoretical Ideas and Future Detectors

Solar Monopoles and Terrestrial Neutrinos
Joshua Frieman — 187

Supernova Neutrino Dynamics: What We Have Not
Learned from Supernova 1987a and What To Expect
from the Coming Galactic Supernova
S.A. Bludman & P.J. Schinder — 194

Ultra-High Energy Neutrino Interactions
and Compositeness
G. Domokos & S. Kovesi-Domokos — 217

New Detectors for Supernova Neutrino Burst Observation
D. Cline — 233

A "New" Old Neutrino-Detector Concept
Peter C. Bosetti — 249

Grande: A Gamma-Ray and Neutrino Detector
H. Sobel et al. — 255

Future Directions for the IMB Detector
J.M. LoSecco — 262

Cosmions and Stars
P. Salati — 271

After Dinner Remarks

"Some Comments on the Problem of Funding Fundamental
Research" and "The Sun Is A Nova"
F. Reines — 285

OBSERVATIONAL NEUTRINO ASTRONOMY

A. NEUTRINOS AND GAMMA RAYS

SUPERNOVA NEUTRINOS

Adam Burrows
Department of Physics and Department of Astronomy
University of Arizona, Tucson, AZ 85721, USA

Abstract. I compare the theory of the neutrino signature of neutron star formation to the Kamiokande II and IMB neutrino data and extract information concerning the nature of the residue in the center of SN1987a.

DISCUSSION

On February 23, 1987, the neutrino burst from the collapse of the core of a massive star was at last detected. To my mind, the neutrino detections alone will mark this Type II as an epochal event to be studied in text books and popular articles for decades to come.

In this paper, I will sketch what we expected the neutrino signature to look like before the explosion and what we may now reasonably conclude from these data. I will not provide detailed and complicated arguments and will not attempt to be uncontroversial, merely credible. There is a danger of overinterpretting these very few events and seeing patterns that are not statistically significant. Nevertheless, the minimum one can extract from these detections is more than adequate to test the basic predictions of stellar collapse and neutron star birth theory. That the characteristics of these neutrino events are consistent with previous predictions is a little surprising. Core collapse has been studied for 30 years and neutron stars for 50 years in blissful theoretical isolation. That the theorists were "on the money" speaks well for the power of scientific speculation and calculation and is a testament to the stalworth pundits of the last half century who nurtured a field that only now is being tested at its core.

The essentials of neutron star structure determine the basics of the neutrino signature of its formation. The canonical neutron star has a radius of 10 kilometers and a gravitational mass of $\sim 1.4\ M_\odot$. The corresponding matter densities range between 10^{14} and 10^{15} gm/cm^3, in the general vicinity of the density of the nucleus. As the name suggests, neutron stars are composed predominantly ($\sim 95\%$) of neutrons, with a small admixture of electrons and protons to ensure beta equilibrium ("$n \leftrightarrows p + e^-$"). There are now probably no less than 10^8 neutron stars in our galaxy, of which $\sim 10^5$ are radio pulsars and ~ 20 are x-ray pulsars. Simple (hopefully, not simple-minded) arguments suggest that $\sim 10^3$ of these collapsed objects are within 10^2 parsecs of the earth.

The small radii and large mass of these objects imply prodigious binding energies of the order of $\sim 0.1-0.2\ M_\odot$. The stellar progenitors of these stellar corpses are only marginally bound with respect to this. Therefore, to form a neutron star, energy must be radiated. The standard theory requires that the lion's share of this energy come off in neutrinos. Neutrinos are not only copiously produced at the high temperatures and densities achieved subsequent to stellar collapse, but can escape on a timescale that is significantly shorter than that of the photons that are the traditional work horses of stellar evolution. However, the high neutrino energies (10-200 MeV) and high densities in the core render the cooling and neutronizing residue of collapse partially opaque to these neutrinos. Neutrinos, therefore, do not stream out of the core within milliseconds, but must diffuse out over the much longer timescale of seconds. The proto-neutron star is a hot, lepton-rich, bloated ember that cools off and shrinks quasi-statically in much the same way as an isolated white or brown dwarf. Its "Kelvin-Helmholtz" evolution can be handled by the standard numerical techniques of stellar evolution theory.[1]

The major sources of opacity are charged-current absorption (involving both ν_e and $\bar{\nu}_e$), electron scattering, and neutral-current scattering off of free neutrons and protons (involving all neutrino species). The initial integrated depth of the star to ν_e's is $\sim 10^4$, whereas that for the mu and tau neutrinos is $\sim 10^3$. The object is a "neutrino star" in the sense that it radiates neutrinos of all species in roughly equal amounts from a "neutrinosphere," just as a regular star radiates photons from a photosphere. The pre-SN1987a list of predictions for the neutrino signal are given in Table 1. Note that $\bar{\nu}_e$'s were predicted to dominate the signal in water detectors due to the large cross-section of the $\bar{\nu}_e + p \rightarrow n + e^+$ reaction and the free proton-richness of water. Note also that the total emitted neutrino energy is expected to dwarf by two orders of magnitude the energy of the supernova explosion itself. From the energetic point of view, the supernova of a sideshow to the main event: neutron star birth. There is not expected to be much signal from the dynamical phase of collapse, rebound, and shock formation. The core energy is trapped on collapse timescales and must be released on neutrino diffusion timescales after the dense core has formed. The average observed ν_e or $\bar{\nu}_e$ energy should be 10-20 MeV, while the average energy of the ν_μ's and ν_τ's (and their antiparticles) should be 25-30 MeV.[2] It is curious and, perhaps, unfortunate that most of the binding energy is expected to be radiated in the non-electron-type species that are more weakly coupled to the matter in detectors than the electron types.

Table 1. Neutrino Signature in Water: Standard Model

* Prompt ν_e burst: milliseconds; $E_{\nu e} < 10^{52}$ ergs (breakout)
* Long-term cooling and neutronization: seconds; $\sim 3 \times 10^{53}$ ergs

  total energy: $\nu_e : \bar{\nu}_e : "\nu_\mu" :: 1.5 : 1 : 4$; all species

  BUT

  total signal: $\nu_e : \bar{\nu}_e : "\nu_\mu" :: 2 : \underline{40} : 1$

  ↑
  $\bar{\nu}_e + p \rightarrow n + e^+$

* Average energy of $\nu_e, \bar{\nu}_e$: ~10-20 MeV; not 100 MeV or 0.1 MeV
* "Neutrino star" embedded in core of massive star
* $\boxed{3 \times 10^{53} \text{ ergs}} \gg 10^{51} \text{ ergs} \gg 10^{49} \text{ ergs}$

  ↑   ↑   ↑
  (ν's)   (SN K.E.)   (light)

The few, but precious, neutrino data from both the Kamiokande (KII) and IMB collaboration[3,4] are shown in Table 2. Not in the table is the important fact that the neutrinos were observed before the light (Figure 1). Collapse is supposed to initiate the sequence of events that leads to the disassembly of the star and the optical pyrotechnics. Had it been the other way around, there would have been an unusual number of puzzled (and skeptical) faces in the corridors of many a physics and astronomy department.

Table 2. Data from the Kamiokande II and IMB detectors.

Event	Time (sec)	Electron energy (MeV)	Angle with respect to Large Magellanic Cloud (degrees)
Kamiokande II			
1	0.000*	20.0±2.9	18±18
2	0.107	13.5±3.2	15±27
3	0.303	7.5±2.0	108±32
4	0.324	9.2±2.7	70±30
5	0.507	12.8±2.9	135±23
6**	0.686	6.3±1.7	68±77
7	1.541	35.4±8.0	32±16
8	1.728	21.0±4.2	30±18
9	1.915	19.8±3.2	38±22
10	9.219	8.6±2.7	122±30
11	10.433	13.0±2.6	49±26
12	12.439	8.9±1.9	91±39
IMB			
1	0.00*	38 (±25%)	74 (±15)
2	0.42	37 "	52 "
3	0.65	40 "	56 "
4	1.15	35 "	63 "
5	1.57	29 "	40 "
6	2.69	37 "	52 "
7	5.59	20 "	39 "
8	5.59	24 "	102 "

*Time for initial event set equal to zero.
**Excluded by Kamiokande II collaboration.

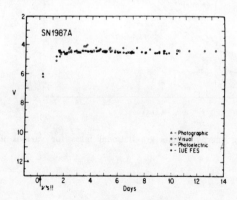

Fig. 1. A compilation of V magnitude estimates for SN1987a in the first weeks of the event. The time of the neutrino detections by KII and IMB is indicated in the lower left.

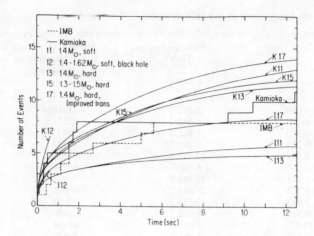

Figure 2. The integrated signals versus time in KII and IMB, as obtained from some model calculations, if the LMC supernova is 50 kpc away, using the techniques in Ref. 1. The actual KII and IMB data are superposed for comparison. No attempt has been made to fit the data. See the text for a more detailed discussion.

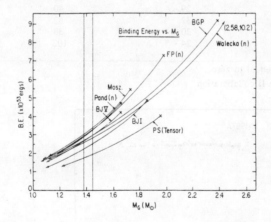

Figure 3. The binding energy versus gravitational mass for various neutron star equations of state.

The large angles with respect to the LMC in Table 2 indicate that, indeed, most of the events were $\bar{\nu}_e$ events, since they lead to a roughly isotropic distribution of detectable positrons. Electron neutrino scattering events are much more forward-peaked. The first one or two events may be ν_e events, but the statistics are such that his can never be proven. It need not be the case. Pre-SN1987a calculations would have predicted a total of no more than one ν_e event in either detector.[5] Perhaps the most important fact that can be extracted from Table 2 is that the signal lasted for seconds, not milliseconds. Some (too many to reference) have suggested that this long duration indicates that the neutrino must have a mass. A nonzero neutrino mass and a spectrum of energies will result in a spread of arrival times if the intrinsic burst duration at the source is short. However, as I have stated, the long signal duration is a predicted and straightforward consequence of the large opacity of the residue to neutrinos. The signal in KII at 9-12 seconds is easy to fit without a mass. Furthermore, much has been made of the seven-second gap in the KII data. Pulses at the source, neutrino masses, and neutrino oscillations, and combinations thereof, have all been evoked. To my mind, the best explanation is that the gap is simply a statistical fluctuation. Not only do the IMB data fill the gap, but small number statistics suggest that with a total of 11 events, gaps are expected.[6] An a priori probability calculation with a reasonable underlying distribution function reveals that the probability of some sort of bunching and gaps of seconds is quiet large. Many workers have been fooled by the low probabilities of a posteriori calculations. Since the stellar evolution calculations[1] predict steady signal out to tens of seconds, Occam's razor suggests that, on occassion, one can be both conservative and correct.

An analysis of the detected energies in Table 2 leads one to derive an average temperature of 3 MeV and 4.5 MeV from the KII and IMB data, respectively.[7] These are well within the estimates of the community prior to SN1987a.[8,9] The IMB detector, due to its high threshold (~20 MeV), is a better thermometer than KII ($\bar{\nu}_e$ threshold ~ 7 MeV). Higher temperatures are required to get any signal at all in IMB. Hence, 4-5 MeV can be shown to be a better estimate of the average underlying temperature of the $\bar{\nu}_e$ neutrinosphere. Data, models, the distance to the LMC (~50 kpc), and the above estimated average temperature yield a total $\bar{\nu}_e$ energy of $3-5 \times 10^{52}$ ergs. Since theory implies that ~1/6 of the total binding energy comes off in $\bar{\nu}_e$'s, we obtain $\sim 2-3 \times 10^{53}$ ergs for the inferred total energy radiated. This is gratifyingly close to the binding energy of a neutron star and is our first measurement of this quantity.

The results of model calculations of the expected signal in KII and IMB are provided for comparison to the actual KII and IMB data in Figure 2. Depicted are the integrated signals in the detector, with the required corrections for the detector efficiencies and thresholds. These particular models were done with and without further accretion and with either a soft or stiff nuclear equation of state, as indicated in the figure legend. The I prefix indicates IMB and the K prefix indicates Kamiokande II. Note that in model 12, a black hole formed. The reader is referred to Ref. 10 for a more detailed discussion of model calculations. The series of calculations that the subset in Figure 2 represent imply that the baryon mass of the neutron star formed in the LMC is between 1.2 $M_\odot$ and 1.7 $M_\odot$, with a preferred range of 1.3 -1.5 $M_\odot$. As Figure 2 indicates, the data do not look like a black hole formation signature. Though a medium-to-small size neutron star fits the data, the poor statistics do not allow us to eliminate larger cores out of hand. Figure 3 shows the run of binding energy versus gravitational mass for various neutron star equations of state not yet eliminated by other considerations. The range of possibilities, not the specific numbers or EOS's, is what is relevant. As Figure 3 indicates, the inferred binding energy implies that a gravitational mass greater than ~1.7 $M_\odot$ is unlikely. As mentioned above, unlikely, also, is that a black hole formed. Black hole formation, when the core is not rotating rapidly (period < 5 milliseconds), looks like neutron star formation and must involve an intermediate proto-neutron star state. This proto-neutron star would collapse dynamically (<1 millisecond) to a black hole only after it accreted a critical neutron star mass. Stellar mass black hole formation should have a neutrino signature that is qualitatively similar to that of neutron star formation but lasts a shorter time. This signal is abruptly cut off when the general relativistic instability sets in. The effect of very rapid rotation on the neutrino signature remains unclear and should be investigated. However, there are indications from analyses of radio pulsar birth statistics[11,12] that most pulsars are born rotating slowly, with initial periods of perhaps hundreds of milliseconds. If the residue of SN1987a has such a long period, the spherically symmetric neutrino calculations will more than suffice.

In sum, the standard theory of the neutrino signature of neutron star birth seems to have passed all the simple tests. However, we should note that the residue of the progenitor of SN1987a has not actually been seen or been directly identified. Nor has a connection between the dense core and the "mystery spot" or "companion" been ruled out. What actually will be revealed when the ejectum disperses? The supernova is evolving so quickly that when you read this, you may know the answer.

ACKNOWLEDGMENT

I would like to thank the Alfred P. Sloan Foundation for its generous support and J. M. Lattimer for his many years of active collaboration.

REFERENCES

1. A. Burrows and J. M. Lattimer, Ap. J. 307, 178 (1986).

2. J. R. Wilson, R. Mayle, S. E. Woosley, and T. A. Weaver, in Proceedings of the 12th Texas Symposium on Relativistic Astrophysics, ed. M. Livio and G. Shaviv, Ann. N.Y. Acad. Sci. 470, 267 (1986).

3. K. Hirata et al., Phys. Rev. Lett. 58, 1490 (1987), (Kamiokande II collaboration).

4. R. M. Bionta et al., 1987, Phys. Rev. Lett. 58, 1494 (1987), (IMB Collaboration).

5. A. Burrows and T. J. Mazurek, Ap. J. 259, 330 (1982).

6. J. M. Lattimer, in Proceedings of the Minnesota conference on SN1987a, eds. T. Walsh and K. Olive, Minneapolis, Minn., June 4-6, 1987.

7. A. Burrows and J. M. Lattimer, Ap. J. (Letters), 318, L57 (1987).

8. S. E. Woosley, J. R. Wilson, and R. Mayle, Ap. J. 302, 19 (1986).

9. R. Mayle, J. R. Wilson, and D. N. Schramm, Ap. J. 318, 288 (1987).

10. A. Burrows in the Proceedings of the George Mason Workshop on SN1987a, Oct. 12-14, 1987, Fairfax, VA.

11. R. Narayan, Ap. J. 319, 162 (1987).

12. R. Chevalier and R. T. Emmering, Ap. J. 304, 140 (1986).

The Nineteen Neutrinos of SN 1987a: Mass or No Mass?

Terry P. Walker
Department of Physics
Boston University
Boston, MA 02215

Abstract

Neutrino mass limits based upon the 19 events from SN 1987a are presented. The correct statistical analysis for sparse data sets, the maximum likelihood method, is reviewed in the context of the SN 1987a neutrinos and the neutrino mass information they contain. Using all 19 events and a reasonable model for SN 1987a neutrino emission, we find a 95% confidence level upper limit to the $\bar{\nu}_e$ mass of 8 eV.

The fact that two large water Čerenkov detectors, operated by the Kamiokande[1] and IMB collaborations[2], detected neutrinos from a supernova in the Large Magellanic Cloud (SN 1987a) has provided dramatic evidence that we understand the generic mechanism responsible for the formation of Type II supernovae: gravitational collapse of a massive stellar core into a neutron star or black hole. In such a scenario, the gravitational binding energy generated during collapse is a few times 10^{53}erg and is most easily liberated by the emission of neutrinos[3]. Maximum likelihood analyses of the neutrino events (11 in Kamiokande and 8 in IMB) seem to show that SN 1987a was indeed the result of such a gravitational collapse, and that the observed neutrinos were trapped on time scales of a few seconds as they thermally radiated with a characteristic temperature of 3-6 MeV[4].

With such remarkable agreement between simple theory and experiment, it is not surprising that astrophysicists and physicists alike have been seduced into trying to milk the sparse neutrino data from SN 1987a for all it is worth. Their goal has been to pull out *detailed* features of both supernova and neutrino physics from the 19 neutrino events. Such attempts are to be viewed with caution for two reasons. The first flag attracting the suspecting eye is the fact that the observation of neutrinos from SN 1987a, few though they may be, implies that Type II supernovae are *dominated* by neutrino emission (the light emitted and kinetic energy of the shock represent only a few percent of the total energy released and gravitational radiation even less). Unlike the ultra-solar-parameter sensitive ^{8}B-solar neutrino flux long sought by Ray Davis with his ^{37}Cl detector, one would expect the neutrino flux from SN 1987a to be rather insensitve to small changes in the parameters which characterize collapse and shock dynamics. Even a supernova which generated a few hundred times more events than 1987a could prove a poor diagnostic for the variations in tens-of-millisecond structure expected in various detailed models of gravitational collapse[5].

The second reason for caution lies with the danger of the statistics of a few events. I choose to illustrate the perils of anemic data by discussing how one extracts information about neutrino masses from SN 1987a. In order to probe the neutrino mass, many authors[6] have attempted to use the fact[7] that a neutrino having mass m which is observed at a time t_{obs} with energy E must have been emitted from SN 1987a at a time t_{em} given by the expression

$$t_{em} = t_{obs} - \frac{D}{2c}(\frac{m}{E})^2, \tag{.1}$$

where D is the distance to SN 1987a (55 kpc $\pm$ 15 kpc[8]) and I have omitted an irrelevant constant. The results of these attempts fall into two categories:

1. By comparing t_{obs} and E of all the events, *exact* values for 1, 2, or 3 distinct neutrino masses, ranging from a few eV to a few tens of eV, are obtained.

2. By examining the resultant spread in emission times (obtained by applying eqn(1) to the observed data) and demanding that it does not exceed what a supernova ought to do, *upper limits* to the neutrino mass result.

Which of these results are we to believe? The first type of analyses suffer from the malady of statistically insignificant peaks in sparse data. Seemingly harmless simplifying approximations, such as neglecting the errors in event energies, can lead to erroneous neutrino masses. The second set of analyses are, by definition, more reliable than the first set. However, they do not involve the correct method of the statistical analysis required when dealing with a few events and therefore cannot assign confidence levels to quoted mass limits. The correct statistical analysis to be used in these neutrino mass studies is the method of maximum likelihood.

Given a model which predicts the emission times and energies of neutrinos from SN 1987a, denoted symbolically by the function f, the likelihood function, L, is simply the product of the relative probabilities of the occurence of each detected event (*i.e.*, the relative probability of the occurence of a set of data with respect to the model f):

$$L \propto \prod_i^{events} \frac{1}{\sigma_i} \int dE \epsilon(E) \sigma(E) e^{-\frac{1}{2}\left(\frac{E-E_i}{\sigma_i}\right)^2} f(t_i \to t_{em}; m^2). \qquad (.2)$$

Here $\epsilon(E)$ is the efficiency of the detector, $\sigma(E) \propto p_e E_e$ is the $\overline{\nu_e}p \to ne^+$ cross section (assuming all the events are of this type[9]), and the integration over dE gaussian smears the energies of all the events weighted by the quoted 1-σ errors σ_i. The function f is a parameterization of the way SN 1987a spits neutrinos. In particular, the emission times are obtained from the arrival times and energies via eqn.(1) and f then assigns a relative probability that SN 1987a would produce the set of $\{t_{em}, E\}$ which result for each m^2. In general, the mass limits obtained from the maximum likelihood method will be somewhat model dependent. However, the sparse neutrino data cannot descriminate between models which well describe the properties expected of the neutrino flux from supernova. The models used for f should contain the relative off-set between the first events of each experiment[10] and other free parameters which describe the rise and decay times of the neutrino emission as a function of energy. For f we choose a diffusion model and vary the functional dependence of the diffusiion coefficient on energy. The KS measures for our models show that they can describe the data adequately. Analyses done in this fashion set the 95% confidence level limit to the $\bar{\nu}_e$ mass at between 10 and 15 eV[11,12] and a similar limit has been obtained using the Kamiokande data alone[13].

Are there any hidden skeletons in the maximum likelihood closet? One might worry about the relative importance of a given event in establishing a mass limit. From eqn.(1), we see that it is the lowest energy events which give the best constraint on the neutrino mass. The lower energy events also carry the most weight in the sample since they are more difficult to see due to the relatively small detection cross section and efficiency. For the same reasons, such events have a greater chance of being background. To this end, the authors of Ref.[11] removed the lowest energy event from the Kamiokande sample and as just mentioned, the 95% confidence level mass limit inflates to $\sim$ 28 eV. It is not kosher however to simply erase the lowest energy event from the data set. This event after all was recorded and has some probability ($\sim$ 75%) of being real. The thing to do is add the expected background to the data set and then examine the mass limits. Unfortunately, we will have to wait for the experimentalists to understand and release this information before making use of it.

Is it possible to extract any particle physics from the 19 neutrino events of SN 1987a? In the case of constraints derived from energy budget arguments, the answer is obviously yes. For the case of constraints involving the statistics of a few events, like the $\bar{\nu}_e$-mass, I believe so. Using the correct type of statistical analysis along with very reasonable models for supernova physics, it appears that SN 1987a yields mass limits which are competitive with present day Tritium-endpoint results. For the experimenters who may feel the competition is a bit too stiff, solice may exist in the form of misunderstood backgrounds or better understood supernova models. In any event, the 19 neutrinos of SN 1987a, with their expected information about neutrino mass and unexpected information on particle physics, are certain to influence the neutrino physics community as they prepare for our next supernova.

References

1. K. Hirata et al.(Kamiokande II Collaboration), *Phys. Rev. Letters* **58**, 1490(1987).

2. R.M. Bionta et al.(IMB Collaboration), *Phys. Rev. Letters* **58**, 1494(1987).

3. S.A. Colgate and R.H. White, *Ap. J.* **143**, 626(1966).

4. A. De Rújula, *Phys. Lett.* **193B**, 514(1987); J.N. Bahcall, T. Piran, W.H. Press, and D.N. Spergel, *Nature*, **327**, 682(1987); A. Burrows and J.M. Lattimer, Arizona Theoretical Astrophysics Preprint # 87-10(1987); D.N. Schramm, Fermilab-Pub-87/91-A(1987); L. Krauss, Yale Preprint(1987); D.N. Spergel et al., Institute for Advanced Study Preprint(1987); S.A. Bludman and P.J. Schinder, University of Pennsylvania Preprint, UPR-0336T(1987).

5. See for example, R. Mayle, J. Wilson, and D.N. Schramm, *Ap. J.*, in press(1987); A. Burrows and J.M. Lattimer, *Ap. J.* **307**, 178(1986); S.E. Woosley, J. Wilson, and R. Mayle, *Ap. J.* **302**, 19(1986); W.D. Arnett, *Ap. J.* **263**, L55(1982); R. Mayle and J. Wilson, Lawrence Livermore National Laboratory Preprint(1987).

6. Rather than construct a surely incomplete list (at last count, there have been at least 40 papers written on SN 1987a neutrino masses), I refer the reader to 3 published papers and the references they contain: J. Bahcall and S.L. Glashow, *Nature* **326**, 476(1987); W.D. Arnett and J. Rosner, *Phys. Rev. Letters* **58**, 1906(1987); E.W. Kolb, A.J. Stebbins, and M.S. Turner, *Phys. Rev.* **D35**, 3590(1987).

7. G.I. Zatsepin, *JETP Lett.* **8**, 205(1968).

8. See E.W. Kolb et al., of Ref.[6] for discussion and further references.

9. The forwardness of the first two events of the Kamiokande is consistent with $\nu e \to \nu e$ scattering. Typical expectations are that isotropic events should outnumber forward events roughly 10 to 1. It cannot be determined with any reasonable confidence that these are scattering events from the very early phases of gravitational collapse (see Ref.[4]). Therefore, I do not discuss any neutrino mass limits which depend on this assumption. Excluding these events from the this analysis of neutrino mass has little effect on the mass limits.

10. The absolute times of the Kamiokande events are believed to be in error by ± 1 minute due to a post-supernova power failure in their mine and therefore it is necessary to include the off-set as a free parameter.

11. L.F. Abbott, A. De Rújula, and T.P. Walker, Boston University Preprint BUHEP-87-24(1987); CERN-TH-4799/87(1987), *Nucl. Phys. B*, in press.

12. D.N. Spergel and J.N. Bahcall, Institue for Advanced Study Preprint(1987), *Phys. Letters* **B**, in press.

13. A. Burrows, in preparation(1987).

COLD DARK MATTER:
A SOURCE OF ENERGETIC PARTICLES

Joseph Silk
University of California
Berkeley, California 94720

> Non-baryonic matter is a generic relic of the early universe which, in the case of cold dark matter, provides a successful theory for large-scale structure. Its inevitable presence in galaxy halos generates a potentially rich signature if the cold dark matter consists of a massive weakly interacting particle species. Cosmology specifies the generic annihilation rate, and energetic byproducts of annihilations that are potentially detectable include antiprotons and postitrons trapped in the halo, gamma rays from the galactic bulge, and neutrinos from the sun.

1. INTRODUCTION

In this talk I am going to discuss a potential source of energetic neutrinos and other particles that is far more speculative than the birth of a supernova. There are enormous uncertainties in estimating the fluxes that I will describe, dominated above all by the possibility that the source itself may not exist. I will focus on the astrophysical uncertainties in the nature of, amount of, and detectable fluxes from, annihilations of cold dark matter candidates. The following speaker will describe some specific particle physics models for particular dark mass particles and the ensuing annihilation spectra.

Cold dark matter is the cosmological definition of a generic, stable, weakly interacting dark matter candidate, which has been non-relativistic since T >> 1 MeV. Most candidates (in particular, those I will discuss here) have masses m_x in the 1-100 GeV range; the invisible axion is one exception. Popular candidates include the least massive supersymmetric partner of known particles: this may be the photino, higgsino, or gravitino in current models. Since weak interactions play a negligible role in thermal equilibrium at $T < 1$ MeV, any remnant cold dark matter, which inevitably was thermally produced in the Big Bang at $T \gtrsim m_x$, must have negligible velocity dispersion in the matter-dominated era $T \lesssim 10$ eV. Hence during this epoch of galaxy and galaxy cluster formation, cold dark matter clusters freely on all scales of astrophysical interest.

Remarkably, one can calculate precisely how much cold dark matter survives today, if one knows the particle properties. The problem, of course, is that we have no evidence at present that such particles exist. Lack of evidence has never deterred theorists, however, and in fact has motivated an alternative approach. This goes as follows. Let us assume that cold dark matter exists in a specified amount: then one appeals to cosmology to infer the interaction strength of the dark matter candidate. This "hands-off" approach leads to specific predictions for observing dark matter candidates, and, in particular, for designing experiments.

The argument goes as follows (Zel'dovich 1965; Chiu 1965). At $T > m_x$, the number density of x-particles is $n_x \sim n_\gamma$. As the temperature drops below m_x, x-particle creation gradually ceases, and the number density is suppressed by a Boltzmann factor:

$$n_x \sim n_\gamma y^{3/2} \exp(-y), \tag{1}$$

where $y \equiv m_x/T$. Expression (1) becomes invalid once the annihilation rate becomes slower than the Hubble expansion rate. Equation (1) tells us that this occurs at a temperature which depends very weakly (logarithmically) on the particle annihilation cross-section. For typical weak interaction rates, one expects $\langle\sigma v\rangle_{ann} \sim 10^{-26}$ cm^{-3}s^{-1}. One finds that freeze-out (fixed comoving density n_x) occurs at $y_f \simeq 20$ or $T_f \equiv m_x/20$.

The number of cold relics is given by equating the annihilation rate to the expansion rate at T_f, or

$$n_x \langle\sigma v\rangle_{ann} \sim H^2. \tag{2}$$

Since $n_x \propto T^3$ and $H \propto T^2$, this yields $\rho_x \equiv n_x m_x \propto \langle\sigma v\rangle_{ann}^{-1}$. A precise value is, measuring ρ_x in units of the critical density for closure of the universe $8H_0^2/8\pi G$, where $H_0 \equiv 100h$ km s^{-1}Mpc^{-1}, $0.5 \leq h \leq 1$, is Hubble's constant,

$$\Omega_x = 0.96 \langle\sigma v\rangle_{26}^{-1} h^{-2} y_{20} f. \tag{3}$$

Here $\langle\sigma v\rangle_{26} \equiv \langle\sigma v\rangle_{ann}/10^{-26}$cm^{3}s^{-1}, $y_{20} \equiv (y_f/20)$ and $f \simeq 1$ is a factor which includes a temperature correction to the annihilation cross-section and uncertainty on the number of relativistic species present at freeze-out. It is a completely unexpected coincidence that for typical weak interaction cross-sections ($\sigma \sim 10^{-36}$cm^2), the density of surviving relics with mass in the 1 - 10 GeV range yields $\Omega_x \sim 0(1)$ (Dicus et al. 1977, Hut 1977; Lee and Weinberg 1977; Sato and Kobayashi 1977; Vysotskii et al. 1977).

In the remainder of this presentation, I will review the evidence for Ω_x, and describe how, given Ω_x, one may apply (3) to predict observable consequences of CDM annihilation occurring today.

2. EVIDENCE FOR DARK MATTER

The evidence for non-baryonic dark matter rests on three arguments, no one of which is compelling, but which collectively provide a strong case.

A. The standard cosmological model of the Big Bang requires $\Omega = 1 + Kt/t_0$, where t_0 is the present epoch ($t_0 = 2/3H_0$ if $\Omega = 1$) and K, a measure of curvature, is observationally bounded by $-0.9 \leq K \leq 1$. Inflation results in a flat universe with $\mid K \mid \lesssim 10^{-4}$. The amplitude of K is fitted empirically by normalization of the inflation–amplified quantum fluctuations to the observed large–scale structure, together with the inflationary prediction of a scale–invariant gaussian fluctuation spectrum. Inflation does not require that Ω consists of non–baryonic matter. However complementary arguments do constrain the baryonic component.

B. The microwave background is uniform, over angular scales from $1'$ to $90°$, to $\delta T/T < 10^{-4}$. If large–scale structure has developed by gravitational instability from primordial adiabatic fluctuations, there should be temperature fluctuations induced at last scattering (when $z \sim 1000$). The fact that these are not seen at a level $\delta T/T \sim 10^{-4}$ suffices to rule out a baryon–dominated universe, since baryonic fluctuations only undergo growth by a factor $\delta\rho/\rho \propto (1+z)^{-1}$ on galactic scales after matter–radiation decoupling at $z \sim 1000$. One would expect $\delta T/T \sim 10^{-3}$ in order for galaxies to have formed by $z \sim 5$, say. In a non–baryonic matter–dominated universe, fluctuation growth is initiated much earlier, at the epoch of matter–radiation equality $1 + z = 4\times10^4\Omega h^2$, if $\Omega \sim 1$. There is also weak (logarithmic) growth in the radiation–dominated era on subgalactic scales. The net effect is that one predicts $\delta T/T \sim 10^{-5}(\Omega h)^{-1}$. Reducing Ω limits the period over which gravity can aid fluctuation growth to $1 + z \lesssim \Omega^{-1}$. The microwave background uniformity therefore requires non–baryonic matter with $\Omega h \gtrsim 0.2$ This lower bound is valid for adiabatic fluctuations, if we adopt the inflationary prediction of a scale–invariant spectrum. A steeper spectrum weakens this limit; however, a purely baryonic universe containing adiabatic fluctuations is untenable for *any* power–law fluctuation power spectrum.

C. Primordial synthesis of ^{4}He, ^{2}H, ^{3}He, and ^{7}Li directly involves the baryonic components and requires $\Omega_b h^2 \simeq 0.03$ to within about a factor of two in order for consistency with the observations of these light elements. Ω_b is bounded both from below, otherwise excessive ^{7}Li is produced, and from above, in order to synthesize enough ^{2}H and not overproduce ^{4}He.

Arguments A and/or B, together with C, imply that $\Omega > 0.1$, and indeed may be of order unity. Unfortunately, direct determinations of Ω by dynamical measurements do not provide any unambiguous support for $\Omega > 0.1$. One can summarize these determinations in terms of the ratio of mass, measured by spectroscopic determinations of velocity, to blue-band luminosity within a specified scale, both measured in solar units. One finds that a distribution of old stars, characteristic of the spheroid of a spiral or an elliptical galaxy, weighs in at $M/L \simeq 10h$. Spirals typically have $M/L \sim 3h$ within their luminous confines. The relevant scale is about 10 kpc for these measurements. Dark halos of spirals, probed by rotation curves, extend to at least ~ 50 kpc, within which $M/L \simeq 50h$. The largest M/L values are measured for groups and clusters of galaxies over ~ 1 Mpc, and for superclusters, over ~ 10 Mpc. Both statistical measures of clustering and galaxy peculiar velocities probe scales up to ~ 50 Mpc. One can say with some certainty that between ~ 1 Mpc and ~ 10 Mpc, the measured M/L is about $300h$. On larger scales, it is very uncertain.

This result may be compared with the closure value, inferred from the luminosity density of all luminous galaxies, which can be expressed as

$$M/L = 1500 \, \Omega h.$$

We conclude that luminous matter in ordinary stars amounts to about 0.007 of the critical value, whereas all dynamically measured matter accounts for $\Omega \simeq 0.2$. Dark matter therefore amounts to about 30 times the luminous stellar (and gas) content of galaxies, and could be 150 times larger if $\Omega = 1$.

3. IS THE DARK MATTER NON-BARYONIC?

All astronomical measurements are presently consistent with $\Omega \simeq 0.1$. Two recent results which report that $\Omega \simeq 1$ are subject to unknown amounts of systematic error. These results are based on measurement of the overdensity in a sample of IRAS galaxies extending to ~ 100 Mpc, and on determination of the volume element by obtaining crude redshifts for galaxies out to $z \sim 0.7$. The problem with the first test is that the IRAS sample is biased towards star-forming field galaxies and biased against the most nearby (and therefore extended) galaxies that are expected to dominate any signal; the problem with the second test is that adequate calibration of the redshift technique has not yet been possible. Work in progress should improve the accuracy of these determinations of Ω.

Up to 90 percent of the dynamically determined Ω may be dark baryonic matter. Primordial nucleosynthesis overproduces ^{7}Li unless $\Omega_b > 0.03 h^{-2}$, whereas $\Omega_{luminous} \simeq 0.01$. Acceptance of a Big Bang origin for ^{7}Li requires there to be at least three times more dark than luminous baryons. Astrophysical candidates for dark baryonic matter include white dwarfs, brown dwarfs, neutron stars and black holes. One cannot predict the abundance of any of these objects with any degree of certainty. White dwarfs are only observable for $\sim 10^{10}$ yr, after which they are black dwarfs. Prolific neutron star or white dwarf production would almost certainly be accompanied by excessive heavy element synthesis. The remaining options, black holes, of mass $\gtrsim 100 M_\odot$, or brown dwarfs, of mass $< 0.08 M_\odot$, the hydrogen-burning threshold, are more difficult to constrain. One usually resorts to plausibility arguments: e.g., if even one percent of the matter forming these high or low mass stars were to form solar mass stars, one would not be producing sufficiently black dark matter, or if the efficiency of forming these objects out of primordial gas clouds were fifty percent or less, one would have excessive amounts of gas remaining behind to be detectable.

Unfortunately, cold dark matter does not provide a completely secure alternative to baryonic dark matter. One can, in principle, predict its abundance. However the compatibility of a cold dark matter-dominated universe with observations needs to be critically reassessed in view of recent developments. Cold dark matter has had great success in leading to a theory of galaxy halo formation and galaxy clustering. Problems arise on larger scales, however. In particular one has to confront four distinct observations, confirmation of any one of which would be fatal for cold dark matter.

A. Dark halos around spiral galaxies may terminate at ~ 50 kpc according to studies of the dwarf satellites of the Milky Way galaxy and of lensing distortions of distant field galaxies that are produced by intervening dark halos.

B. The clustering of galaxy clusters amounts to an enhancement of the strength of cluster correlations relative to galaxy–galaxy correlations by a factor of at least 20 on scales of 10 – 20 Mpc.

C. Large-scale bulk motions have been reported, averaged over scales of up to 50 Mpc, that amount to ~ 600 km s^{-1} in the reference frame of the cosmic microwave background radiation. This is in addition to the motion of the Local Group with similar amplitude, inferred from the dipole anisotropy of the background radiation.

D. Temperature fluctuations have been measured in the cosmic microwave background over $8°$ at a strength of about $\delta T/T \simeq 5 \times 10^{-5}$.

None of these observations can yet be considered to be definitive. For example, zero–point variations in the correlation between velocity dispersion and luminosity of giant ellipticals, a correlation for which there is no accepted explanation, would induce distance errors, which would in turn be interpreted, erroneously, as peculiar velocities. However, it is fair to say that the cold dark matter theory makes unambiguous predictions. Normalization of the inflationary scale–invariant, gaussian fluctuation spectrum to the galaxy luminosity function leads to predictions of large–scale power. One infers $\delta\rho/\rho$ as a function of scale, and one can straightforwardly calculate the various observables such as peculiar velocity and $\delta T/T$. Biasing, or appeal to rare fluctuations on large scales, helps accentuate the density contrast and produces large voids, as observed. Indeed biasing on galactic scales is essential to account for the concentration of light relative to the dark matter. However, this effectively lowers the mean amplitude of the dark matter power spectrum relative to that of the luminous matter, thereby compounding the difficulties in producing sufficient large-scale power to obtain observable peculiar velocities and microwave background temperature fluctuations. The value of $\delta T/T$ predicted by unbiased cold dark matter is an order of magnitude below a recently reported value over $\sim 8°$.

However, until definitive confirmation of these various results is forthcoming, cold dark matter merits serious attention because of its many successes on scales below ~ 10 Mpc. These include accounting for galaxy correlations, galaxy groups and clusters, galaxy relative velocities, galaxy rotation velocities, and the structure of galaxy halos.

4. DETECTION OF CDM

Provided that $\Omega_x \gtrsim 0.1$, cold dark matter provides a natural candidate for the content of the dark halo of our galaxy. The dark halo density in the solar neighborhood is known from modelling of the galactic rotation curve to be between 0.03 and 0.3 GeV cm^{-3}. One, therefore, expects a local flux of dark matter particles of at least $10^6 m_x^{-1}$cm^{-2}s^{-1} (with m_x in GeV).

4.1. Laboratory detection

Such a flux is measurable in a laboratory bolometric detector by virtue of elastic scattering with suitable target nuclei. The energy deposition is of order

$$m_x^2 M_{target}^2 v^2 (1 - \cos\theta)(m_x + M_{target})^{-2} \sim 100\text{eV}$$

for a silicon target. The elastic cross-section ($xp \to xp, xn \to xn$) is at least $\sigma_{el} \sim 10^{-38}\text{cm}^2$ for a particle with annihilation cross-section $\sigma_{ann} \sim 10^{-36}\text{cm}^2$, and may be enhanced by as much as a factor of ~ 100 if spin-independent interactions are important (as for heavy neutrinos, for example). The velocity refers to the relative velocity between the earth and the halo particles, and the forward peaking means that seasonal modulation can be important in enhancing a signal.

4.2. Annihilation products

Majorana-mass candidates for cold dark matter annihilate with a cross-section corresponding to the low temperature limit of the known early universe annihilation cross-section, itself known once Ω_x is specified. For generic candidates (photino, higgsino, etc.) other than the scalar neutrino, the annihilation products consist of heavy quarks that decay into $\nu\bar{\nu}, e^+e^-, p, \bar{p}$ and γ rays. To obtain a more precise estimate of their flux and spectrum, one has to specify the dark matter candidate. For example, for the popular supersymmetric candidate, the photino, one has

$$\langle\sigma v\rangle(T_f) \propto m_{sf}^{-4}\{m_f^2 + m_{\tilde{\gamma}}^2\},$$

which, applying (3), yields

$$\Omega h^2 \simeq 0.25(m_{sf}/60\text{GeV})^4(4\text{GeV}/m_{\tilde{\gamma}})^{4/3}.$$

The experimental lower bound $m_{sf} \gtrsim 60$ GeV requires $m_{\tilde{\gamma}} \gtrsim 4$ GeV. Note also that $\langle\sigma v\rangle(T=0) \propto m_{sf}^{-4}$, so that theories which have very massive scalars, as in phenomenological superstring models, result in massive ($\gg 10\text{GeV}$) photinos. For candidates in the 1 – 100 GeV range, the annihilation products are potentially detectable in at least three distinct environments.

4.3. Annihilations in the sun: high energy neutrinos

As the sun orbits the galaxy, it traps halo dark matter. Particles impact the solar surface, the impact rate being boosted by gravitational focussing, and elastic scattering with protons and other nuclei guarantees that they remain and settle in the sun provided that $m_x > 4$GeV. Lighter x-particles evaporate before accumulating in the solar core. The solar trapping rate is

$$\Gamma_T = 7 \times 10^{28} m_x^{-1}(\rho_h/0.3\text{GeV cm}^{-3})v_{300}^{-1}f_E,$$

where $v_{300} \equiv v_x/300$ km s^{-1} is the average velocity of a halo x-particle near the sun and f_E is the trapping probability per incoming particle. According to Srednicki et al. (1987),

$$f_E = (0.18 Y_p \sigma_{el,p}/10^{-36} \text{ cm}^2) \min\{1, 43 m_p m_x (m_p^2 + m_x^2)^{-1} v_{300}^2\}$$

allows for incoming particles to be scattered and energetically trapped.

The equilibrium abundance is inferred by equating annihilation and trapping rates. Annihilations in the solar core yield a neutrino spectrum of which only the high energy neutrinos escape. These may be detectable at the earth given the rapid decline of atmospheric neutrinos with increasing energy. Indeed, Ng et al. (1987) find that the predicted high energy neutrino solar flux of $\sim 10 m_x^{-1}$ cm^{-2}s^{-1}GeV^{-1} for sneutrinos already exceeds IMB bounds, although photinos, producing neutrinos at about one percent of this rate, are not presently constrained.

4.4. Annihilations in the galactic bulge: gamma rays

The gamma ray annihilation luminosity from the galactic halo may be approximated by

$$L_{ann,\gamma} = N_x n_x \langle \sigma_{ann} v \rangle^{-1} \zeta ,$$

where N_x is the total number of halo dark matter particles, n_x is the mean density, and ζ is the gamma ray multiplicity (gammas per annihilation). With N_x corresponding to $10^{12} M_\odot$, one finds $\langle n_x \sigma_{ann} v \rangle^{-1} \simeq 10^{19}$ yr and $L_{ann,\gamma} \equiv 10^{41} \zeta m_x^{-1}$ s^{-1}. At the earth, this amounts to only about one percent of the isotropic gamma ray background above 100 MeV. Anisotropy may help enhance the halo signal; it will also be more significant relative to the background at high energies (above 1 GeV) if $m_x \gtrsim 10$ GeV.

The galactic bulge also provides a potential gamma ray source. One should not necessarily assume that the galactic spheroid (with characteristic r^{-2} density profile, core ~ 10 kpc) is necessarily 100 percent baryonic matter. Formation of the galaxy would inevitably entrain some dark matter, nor do we understand formation so well that one can dismiss the possibility of a dense dark matter core that formed in the early universe around which the galaxy accreted. The bulge of our galaxy contains a mass $\sim 6 \times 10^8 M_\odot$ within a core–radius ~ 100pc, and the inferred mean annihilation time $< n_x \sigma_{ann} v >^{-1} \simeq 5 \times 10^{22}$s leads to a gamma ray luminosity $L_\gamma \simeq 10^{42} \zeta m_x^{-1} f_{b,x}$ s^{-1}, where $f_{b,x}$ is the fraction of the bulge mass in the form of CDM. This is observable if $f_{b,x} \sim 1$, and one can even limit $f_{b,x}$ to $\lesssim 0.1$ using COSB data for some bulge models and dark matter candidates with $m_s \lesssim 10$ GeV.

Unfortunately there is little direct evidence for an enhanced mass–to–light ratio in our own galactic bulge. The situation in M31 is more intriguing, however. Recent high resolution observations of the innermost core of M32 find a high ratio of mass–to–luminosity within the central 2 pc. The inferred mass within this region is $\sim 10^7 M_\odot$. While a central black hole offers one possible explanation, a dark matter core is also viable: the inferred annihilation time is $\langle n_x \sigma_{ann} v \rangle^{-1} \simeq 10^{18}$s, and the gamma ray luminosity is $\sim 10^{46} \zeta m_x^{-1}$s^{-1}. Since the annihilation spectrum of gamma rays turns out to have a different signature from other gamma ray sources, this could lead to an interesting cold dark matter signature, if indeed CDM accumulates in the cores of galaxies.

4.5. Annihilations in the halo: energetic antiprotons

Charged particles produced by annihilations accumulate in the galactic halo, trapped by magnetic confinement for $\sim 10^8$ yr. The rarest stable charged particle annihilation product is the $\bar{p}$, and its production energy is expected to be $\sim 0.2 m_x$. One, therefore, expects a potential cosmic ray $\bar{p}$ signature of CDM (Silk and Srednicki 1984). The only competing source is from high energy (E $\gtrsim$ 7 GeV) protons colliding with interstellar atoms to produce occasional $\bar{p}$ secondaries that are, however, kinematically supressed below ~ 1 GeV. Low energy cosmic ray $\bar{p}$ should therefore provide a clean signature of halo CDM, especially if the data allow sufficient energy resolution to separate the secondary contribution. Experiments flown in August 1987 should at the very least confirm or exclude Buffington's (1981) measurement of low energy $\bar{p}$.

This research has been supported in part by DOE and by CALSPACE.

REFERENCES

Buffington, A. *et al.* 1981, *Ap.J.*,**248**, 1179.
Chiu, H.-Y. 1966, *Phys. Rev. Lett.*,**17**, 712
Dicus, D.A., Kolb, E.W. and Teplitz 1977, *Phys. Rev. Lett.* **39**, 168.
Hagelin, J.S., Ng, K.W., and Olive, K.A 1987, *Phys. Lett.*,**180B**, 375.
Hut, P. 1977, *Phys. Rev. Lett.*,**69B**, 85.
Lee, B.W. and Weinberg, S. 1977, *Phys. Rev. Lett*,**39**, 165.
Sato, K. and Kobayashi, M. 1977, *Prog. Theor. Phys.*,**58**, 1775.
Silk, J. and Srednicki, M. 1984, *Phys. Rev. Lett.*,**53**, 624.
Vysotskii, M.I., Dolgov, A.D. and Zel'dovich, Ya. B. 1977, *JFTP Lett.*,**26**, 188.
Zel'dovich, Ya. B. 1965, *Adv. Astron. Astrophys.*,**3**, 241.

COSMIC-RAY NEUTRINOS AND OTHER STUFF (ANTIPROTONS, GAMMA RAYS AND POSITRONS) FROM GALACTIC DARK MATTER ANNIHILATION*

F. W. Stecker

Theory Group, Laboratory for High Energy Astrophysics
NASA Goddard Space Flight Center
Greenbelt, MD 20771, USA

If Majorana fermions of mass M_X ~15-20 GeV make up the dark matter in the galactic halo, their annihilation products can account for the entire spectrum of observed cosmic-ray antiprotons. Under this assumption, the spectra of the various potentially observable annihilation products of cosmic ray energies, viz., neutrinos, γ-rays, and positrons as well as antiprotons, have characteristic spectra with high energy cutoffs at energy M_X. The flux of annihilation positrons may be comparable to the measured cosmic-ray positron flux in the 1-10 GeV range and may also be comparable to that expected from the decay of collisionally generated cosmic-ray pions (up to energy M_X.) The calculated γ-ray flux from dark matter annihilation in the galactic halo is low compared with cosmic-ray produced γ-ray fluxes at high galactic latitudes, however, its characteristic spectral cutoff may be observable. Annihilation neutrinos and γ-rays from the galactic center may also be observable, the neutrinos above the atmospheric background.

I. INTRODUCTION

The universe is quite a bit heavier than it looks. The general evidence for this has been discussed by Silk in the previous talk (these proceedings; see also Ref. 1.) The observation of flat galactic rotation curves strongly suggests the presence of cold dark matter comprising most of the mass in galactic halos which could also dominate the mass of the universe. Numerical simulations of galaxy formation and distribution indicate that at least a significant fraction of the nonvisible mass of the universe is in this non-baryonic form. Particle physics theory suggests a number of possible candidates for cold dark matter. Among these are neutral, heavy (several GeV) Majorana fermions which are relics of the big bang.* It is, of course, important that the dark matter candidate be stable. The mass eigenstate "neutralino" is generally a superposition of the "higgsino", "photino" and "zino" which are mixed by gauge and supersymmetry breaking[2]. The lightest supersymmetric particle (LSP) is stable by virtue of a natural conservation law, viz., so-called R-parity conservation, since it is the lightest state with odd R-parity. It is most probable that the LSP is either almost a pure higgsino or a pure photino. (Unlike the case with the LSP, there is no natural way of forbidding the decay of a "conventional" heavy Majorana neutrino. For this reason, one could prefer the label "higgsino" to "massive Majorana neutrino" for the ideal dark matter candidate).

A number of papers have appeared suggesting methods for the indirect detection of dark matter particles in our Galaxy. In particular, Stecker,

---

*Existing experimental limits on the fluxes of such neutrinos from deep underground detectors already exclude muon sneutrinos in the mass range of interest as candidate dark matter particles, but do not exclude the Majorana fermion species in this mass range (T. Gaisser, private communication).

Rudaz and Walsh[3] and Rudaz and Stecker[4] suggested that the entire observed spectrum of cosmic-ray antiprotons could be produced by annihilations of such heavy Majorana fermions χ of mass M_χ between about 15 and 20 GeV. The shape of the antiproton spectrum can be reproduced given that range of M_χ, using the experimentally observed single-particle antiproton spectrum from the closely analogous process $e^+e^- \to \bar{p}$ + anything, and taking into account the modulation effects of the solar wind. This annihilation process accounts for an apparent high energy cutoff in the cosmic-ray $\bar{p}$ spectrum around $E_{\bar{p}}$ = 15 to 20 GeV. The general point depends only on the existence of stable particles annihilating by a weak interaction with a nonnegligible branching ratio to quark-antiquark pairs. The quark jets in the final state should then have the same characteristics as those produced in e^+e^- colliders.

In Ref. 3, we presented numerical estimates of the $\bar{p}$ spectrum, assuming the χ particles to be photinos, the spin-1/2 partner of the photon expected in theories with broken supersymmetry. However, to account for the observed magnitude of the spectrum given this assignment, required our choosing values for the masses of the spin-0 partners of the leptons and quarks (sleptons and squarks) which have since then been excluded (in the case of 15-20 GeV photino masses) on the basis of a reanalysis of monojet and dijet data obtained at the CERN $p\bar{p}$ collider[5]. The new bounds on spin-0 partner masses, $M_{\tilde{q}} \geq 60$ GeV, suggest that massive photinos in the halo may not account for the observed $\bar{p}$ spectrum without an extremely long trapping time in the galactic halo. Rudaz and Stecker[4] show that higgsinos or Majorana neutrinos can "fill the bill".

Taking χ to be a generic higgsino (spin 1/2 superpartner of a Higgs scalar in the simplest extension of the standard electroweak model with broken supersymmetry) or, indeed, simply a heavy Majorana neutrino with standard couplings to the Z^0 boson allows us to account for the observed flux and spectrum of the cosmic-ray $\bar{p}$'s[4]. In this case the annihilation cross section is independent of unknown particle physics parameters such as the squark and slepton masses. The relic density of such χ particles where $\chi = \tilde{h}$ or ν_M is easily calculated from well-known formulae to be[6]

$$\Omega_\chi h_{1/2}^2 \simeq 0.2 \ (M_\chi/15 \text{ GeV})^{-2} \qquad (1)$$

a formula valid for $M_\chi \geq 15$ GeV. (Here, Ω is the fraction of closure density and $h_{1/2} = H_0/50$ km sec^{-1}Mpc^{-1}.) Thus, for $M_\chi = 15$ GeV, $\Omega \sim 0.2$, in excellent agreement with the value of Ω associated with galaxies[7], and one which is sufficient to give the halo mass. A species of more diffuse dark matter, e.g., light neutrinos[8], may provide an additional contribution to Ω on a universal scale to bring the total value of Ω up to the value of unity required by inflationary cosmology.

II. THE ANTIPROTON COSMIC RAY SPECTRUM

The production rate of antiprotons from $\chi\chi$ annihilation in the halo is given by

$$Q_{\bar{p}}(E) = n_\chi^2 <\sigma_\chi v>_A f_{\bar{p}}(E) \qquad (2)$$

where $f_{\bar{p}}(E) = dN_{\bar{p}}/dE$, normalized to the number of antiprotons per annihilation. The "cold" annihilation cross-section $<\sigma v>_A$ is overwhelmingly dominated by the contributions of τ leptons, and c and b quarks in the final state. The annihilation cross section is then given by[4]

$$<\sigma_\chi v>_A = \frac{G_F^2}{4\pi} (m_\tau^2 + 3m_c^2 + 3m_b^2) \qquad (3)$$

including a factor of three for color in the case of quarks in the final state and noting that for $M_\chi \geq 15$ GeV, all values of $\beta_f \simeq 1$. Numerically, with $m_\tau = m_c = 1.78$ GeV and $m_b = 3m_c = 5.3$ GeV,

$$<\sigma_\chi v>_A = \frac{31}{4\pi} G_F^2 m_\tau^2 = 1.26 \times 10^{-26} \text{ cm}^3 \text{s}^{-1} . \qquad (4)$$

The antiprotons can only come from the hadronic $c\bar{c}$ and $b\bar{b}$ final states, which in fact account for a fraction $\delta_h \simeq \frac{30}{31} \simeq 1$ of the total. We now require the spectrum $f_{\bar{p}}(E)$ which can be written

$$f_{\bar{p}}(E) = \frac{2}{\sigma_A} \frac{d}{dE} (\sigma_{\chi\chi \to \bar{p} + X}) . \qquad (5)$$

(The factor of 2 takes account of the production of antineutrons which decay to give additional antiprotons.)

This quantity cannot yet be calculated reliably from quantum chromodynamics, but fortunately we can appeal here to experiment in the form of the closely related process, $e^+e^- \to \bar{p}$ + anything. It is an empirical fact that the $\bar{p}$ single particle distribution at all measured initial center-of-mass-energies $\sqrt{s}$ in this process[9] can be fitted by a single function of the form

$$\frac{s}{\beta_{\bar{p}}} \frac{d}{dx}(\sigma_{e^+e^- \to \bar{p} + X}) = 0.4 (8.5 e^{-11x} + 0.25 e^{-2x}) \quad \mu\text{b GeV}^2 \qquad (6)$$

where $x = 2E/\sqrt{s}$ is the scaled antiproton energy. With

$$\sigma_{e^+e^- \to \text{hadrons}} = \frac{4\pi\alpha^2}{3s} R_h = 87 R_h [s(\text{GeV})^2]^{-1} \text{ nb}, \tag{7}$$

and $R_h = 4$ from experiment[10], eq.(6) becomes

$$\frac{1}{\sigma_{e^+e^- \to h}} \frac{d}{dx}(\sigma_{e^+e^- \to \bar{p}+X}) = 1.2\beta_{\bar{p}}(8.5e^{-11x} + 0.25e^{-2x}). \tag{8}$$

The fact that e^+e^- annihilation involves different proportions of lighter quarks than the $\chi\chi$ annihilation with which we are concerned here (where the $b\bar{b}$ final state is overwhelmingly dominant) is not a worry. Experiments with enriched b quark samples indicate very similar single particle spectra as average events, except perhaps for very low values of x (see, e.g., the recently published results of the TPC/Two Gamma Collaboration[11]. Accordingly, we shall take (with $\sigma_A = \sigma_{\chi\chi \to \text{hadrons}}$, $x = E/M_\chi$)

$$\frac{1}{\sigma_A}\frac{d}{dx}(\sigma_{\chi\chi \to \bar{p}X}) = \frac{1}{\sigma_{e^+e^- \to h}}\frac{d}{dx}(\sigma_{e^+e^- \to \bar{p}X}), \tag{9}$$

$$f_{\bar{p}}(E) = 0.16\beta_{\bar{p}}(M_\chi/15 \text{ GeV})^{-1}(8.5e^{-11(E/M_\chi)} + 0.25e^{-2(E/M_\chi)}) \text{ GeV}^{-1}. \tag{10}$$

With the production rate given by eq.(6), the interstellar $\bar{p}$ flux is

$$I_{\bar{p}}(E) = (4\pi)^{-1}Q_{\bar{p}}(E)\beta_{\bar{p}}c\tau = 2.5\times 10^{-5}\beta_{\bar{p}}^2\kappa(8.5e^{-11(E/M_\chi)} + 0.25e^{-2(E/M_\chi)})$$

in $\text{cm}^{-2}\text{s}^{-1}\text{sr}^{-1}\text{GeV}^{-1}$ with \hfill (11)

$$\kappa \equiv (\rho_\chi/0.75 \text{ GeV cm}^{-3})^2 (M_\chi/15 \text{ GeV})^{-3}(\tau/2 \times 10^{15}\text{s})$$

written in terms of ρ_χ, the χ mass density in the halo, with τ being the mean lifetime for escape of antiprotons from the galactic halo.

This interstellar antiproton spectrum from dark matter annihilation, shown as a dashed curve in Figure 1 can be compared with observations only after the modulation effects of the solar wind have been taken into account. The amount of modulation varies with the solar cycle. An effective rigidity dependent radial diffusion coefficient for the time period of the low energy measurement of Buffington and Schlindler[12], i.e., 18 June 1980, can be determined from the Pioneer 10, ISEE-3 and Helios 1 space probe data[13] and used to modulate the theoretical antiproton spectrum[3,14].

Owing to the energy dependence of solar modulation, it is reasonable to estimate the magnitude of the interstellar flux using the highest energy data of Golden, et al.[15], which should be least affected by solar wind modulation. This gives an interstellar $\bar{p}$ flux around 7 GeV (corresponding to x = 0.5) of ~4.6 x 10^{-6} cm^{-2} sec^{-1} sr^{-1} GeV^{-1} modulated to a value of ~2.5 x 10^{-6} in the same units (see Fig. 1) To account for such a flux with $M_\chi \simeq 15$ GeV in eq. (11) requires $\kappa = 1.45 \pm 1.0$, taking account of the experimental errors in both the e^+e^- data and the cosmic-ray antiproton experiments.

It is possible to use dynamical data to estimate the value for the dark halo matter density, ρ_χ, in the solar neighborhood. Ng, Olive and Srednicki[16] have argued that $\rho_\chi \leq 0.75$ GeV cm^{-3}. However, this value is derived for a spherical halo model. The dark matter density in the solar neighborhood may be larger by a factor of ~ 2 to 3 for flattened halo models[17]. Cosmic-ray halo propagation models[18] give lifetimes $\tau \simeq 3 \times 10^{15}$s. Thus, the astrophysical values for ρ_χ and τ are compatible with the requirements derived from the $\bar{p}$ flux.

Using the values for $<\sigma v>_A$ and δ_h given here and in Ref. 3, one can obtain an expression for the ratio of p yields from photino and generic higgsino annihilation,

$$Q_{\bar{p}}^{\tilde{\gamma}} / Q_{\bar{p}}^{\tilde{h}} = [\delta_h^{\tilde{\gamma}} <\sigma_{\tilde{\gamma}} v>_A / \delta_h^{\tilde{h}} <\sigma_{\tilde{h}} v>_A] \simeq 9.4 \times 10^{-2} (m_W/m_{\tilde{q}})^4 \leq 0.32 \qquad (12)$$

where $m_W \simeq 81$ GeV is the W boson mass.

Thus, in order to obtain the observed $\bar{p}$ flux from photino annihilation, we would require a value for κ at least three times larger, given the accelerator limit $m_{\tilde{q}} \geq 60$ GeV. Such a value for κ could be difficult to reconcile with the astrophysically obtained values for ρ_χ and τ unless τ is quite large. Rudaz and Stecker[4] have chosen a normalization to fit the higher energy data. With this chioce, the theoretical curve falls below the data point of Buffington and Schindler[14], although it is still within 2σ of their result (see Fig. 1). Choosing a normalization to provide a best fit to all of the data, would have yielded a satisfactory fit to the low energy point and still been well within the error estimate for κ given above. Note the characteristic prediction of the annihilation spectrum, viz., the existence of a kinematic cut-off in the $\bar{p}$ spectrum above ~ 15 GeV.

III. THE COSMIC RAY NEUTRINO AND POSITRON SPECTRA FROM $\chi\chi$ ANNIHILATION

The cosmic-ray positrons from $\chi\chi$ annihilation provide a source in addition to the flux from the decay of secondary pions produced in cosmic-ray interactions. The positron flux to be expected from the latter process has been well studied and the spectrum is expected to have roughly a power-law fall off with energy in the energy range above 1 GeV. This fact makes a calculation of the spectrum as well as the positron flux quite interesting, particularly for the annihilation channels which give a fairly hard positron spectrum from prompt decay. Of course, at the higher energies, energy losses from synchrotron radiation and Compton interactions must be taken into account when computing the resulting positron spectrum.

In calculating the ν, e^+ and γ-ray fluxes from $\chi\chi$ annihilation, one normalizes to the $\bar{p}$ data of Golden, <u>et al.</u>[15] at $E_{\bar{p}} \simeq 7$ GeV, with modulation taken into account (see above). This approach has the merit of being independent of the various individual astrophysical parameters which determine the absolute intensities of the fluxes, although the implied combination κ takes the value given previously.

There are three sources of neutrinos and positrons from $\chi\chi$ annihilation to consider, <u>viz.</u>, first generation prompt leptons (P1), second generation prompt leptons (P2), and π^+ meson decay leptons (π). We will consider these sources separately.

A. First Generation Prompt Leptons (ν's and e^+'s).

The products of $\chi\chi$ annihilation, τ leptons, and charmed and bottom quarks are efficient sources of prompt, high energy leptons and antileptons from their weak decays. The relevant decay chains are as follows (W* is a virtual W boson):

$$\tau^+ \to \bar{\nu}_\tau + W^{+*}$$
$$\bar{b} \to \bar{c} + W^{+*} \tag{13}$$
$$\text{and } c \to s + W^{+*}$$

with W^{+*} decaying to $\ell^+ \nu_\ell$, where $\ell^+ = e^+$ or μ^+ (for b decays, $\ell = \tau$ is considerably suppressed by phase space and will be neglected here). We will take the branching ratios for a given ℓ^+ type to be respectively 18% for τ (Ref. 19) and 13% for both b and c. Compared to the dominant $b \to c$ and $c \to s$ transitions, the $b \to u$ and $c \to d$ transitions are considerably suppressed and will be neglected here. With $\ell^+ = e^+$, the above decays provide the most energetic, "first generation" prompt positrons.

The normalized positron energy distribution from the first generation decays $\bar{b} \to \bar{c} + e^+ + \nu_e$ and $\tau^+ \to \bar{\nu}_\tau + e^+ + \nu_e$ is, taking final state particles as extremely relativistic, (Note here, $M_f^2/M_\chi^2 \ll 1$.)

$$f(x) = \frac{5}{3} + \frac{4}{3} x^3 - 3x^2 \qquad (14)$$

for the b and τ decay and

$$f(x) = 2 + 4 x^3 - 6x^2 \qquad (15)$$

in the case of $c \to s + e^+ + \nu_e$, where $x \equiv E/M_\chi$, E being the positron energy. Note that M_χ here is the energy of the decaying fermion produced in the $\chi\chi \to f\bar{f}$ annihilations in the excellent relativistic approximation indicated above. The respective average positron energies are 0.35 M_χ and 0.3 M_χ. In what follows, a good approximation to both spectra is provided by the simple step function

$$f(E) = \frac{1}{kM_\chi} \theta (kM_\chi - E) \qquad (16)$$

where $k = 0.7$ gives an average energy of 0.35 M_χ. This approximation to eqs. (14) and (15) will be useful later in the calculation.

It is now easy to write down the first generation positron source function

$$Q_{e^+}(E) = n_x^2 \langle \sigma_x v \rangle_A \{ 0.18 \frac{1}{31} f_\tau(E) + 0.13 [\frac{3}{31} f_c(E) + \frac{27}{31} f_b(E)]\} . \quad (17)$$

Neglecting the difference between the energy distributions and using the approximation (16), eq.(17) reduces to

$$Q_{e^+}(E) = Q_0 \theta(kM_x - E)$$

where (18)

$$Q_0 = 0.135 \, n_x^2 \langle \sigma_x v \rangle_A / (kM_x) .$$

In order to calculate the shape of the cosmic-ray positron spectrum expected to be observed (as opposed to the spectrum at production), it is necessary to consider that the positrons lose energy by synchrotron radiation in the galactic magnetic field and by Compton interactions with photons from starlight and the microwave background radiation.

With galactic positron annihilation negligible for energies above 1 GeV, the resultant steady state equilibrium spectrum is[20]

$$I(E) = \frac{c}{4\pi |r(E)|} \int_E^\infty dE' \, Q(E') \exp(-\frac{1}{\tau} \int_E^{E'} \frac{dE''}{|r(E'')|}) \quad (19)$$

where the energy loss rate from synchrotron radiation and Compton scattering is

$$\dot{r} = -bE^2$$

$$b = \frac{4}{3}(\sigma_T/m_e^2 c^3)(\rho_\gamma + \rho_{mag}) = 1.0 \times 10^{-16} [\rho_\gamma + 0.02\ B^2]\ \text{GeV}^{-1}\text{s}^{-1},$$
(20)

with σ_T being the Thomson cross section, the interstellar photon energy density ρ_γ being given in units of eV cm^{-3}, the mean perpendicular galactic magnetic field strength B being given in μG, and E being given in GeV. Typically, B = 2 - 3 μG in the galactic plane (possibly lower in the galactic halo) and ρ_γ has contributions from the cosmic microwave background radiation (0.25 eV cm^{-3}) and starlight (0.4-0.6 eV cm^{-3} in the galactic plane, again possibly lower in the halo.)

It will be convenient here to define the parameter D such that $(\dot{r}\tau)^{-1} \equiv DE^{-2}$ where, from eq. (20)

$$D = (b\tau)^{-1} = 5\ \text{GeV}\ (\rho_\gamma + 0.02\ B^2)^{-1}(\tau/2 \times 10^{15}\text{s})^{-1}.$$
(21)

With Q(E) given by eq. (18) and $E_0 \equiv kM_x$, $A \equiv cQ_0/4\pi b$, $y \equiv D/E$ and $y_0 = D/E_0$, eq.(19) has the solution

$$I = \frac{A}{D}\{y^2 e^{-y}[\frac{e^{y_0}}{y_0} + Ei(y) - Ei(y_0)] - y\}$$
(22)

where the exponential integral

$$Ei(y) \equiv -\int_{-y}^{\infty} dt\ \frac{e^{-t}}{t}.$$
(23)

It follows from eq.(22) that as E→0 (y→∞),

$$I \to \frac{A}{D}[1+2(\frac{E}{D})]$$

$$\to \frac{A}{D} = Q_0 \tau c/4\pi \tag{24}$$

as expected when energy losses are negligible.

We have solved eq.(19) using the useful closed form solution (22), and also using eqs. (14) and (15) to obtain a more exact solution[4]. Because synchrotron radiation and Compton interactions deplete the high energy tail of the source spectrum, both solutions give essentially the same result.

Using the parameters found to explain the cosmic ray antiproton data, we take E_0 = 10.5 GeV corresponding to M_X = 15 GeV, D = 10 GeV, and the other parameters as given above. The resultant first generation positron spectrum is shown in Fig. 2, labeled (P1). Three characteristics may be noted about the P1 spectrum: (1) it is a constant at the lower energies (see above), (2) there is depletion at the higher energies near the cutoff energy E_0 owing to energy losses, and (3) there is a "pileup" near E = 2.5 GeV, also owing to energy losses from synchrotron radiation and Compton interactions.*

B. Second Generation Prompt Positrons.

Additional positrons come from the second generation prompt processes given by the decay chains

---

*It has been determined that in electron-positron collider events where b and c quark jets are produced, these jets carry off 80% and 60% of the cms energy respectively (Ref. 11). In our calculations, no significant error is expected from neglecting this detail because (1) as indicated by eq. (17), our P1 positrons come overwhelmingly from the b quark channel where 80% of the energy is stll carried away, and (2) synchrotron radiation and Compton losses degrade the high energy tail of the source spectrum preferentially in any case (see eq. (20)).

$$\tau^+ \to \mu^+ \to e^+$$
$$c \to \mu^+ \to e^+ \qquad (25)$$

and

$$\bar{b} \to \mu^+ \to e^+ \,.$$

Another second generation process is the decay chain

$$b \to c \to e^+ \,. \qquad (26)$$

It is easily seen that the branching ratio for the first process in eq. (25) is still 18%, while that of the last three is still 13% (the decay $\mu \to e\nu\nu$ has unit probability, as do $b \to c + W^{-*}$ and $c \to s + W^{+*}$).

The calculation of the second generation spectrum can be performed as follows: All four processes given by eqs. (25) and (26) will be treated in a similar way. First, a muon or charmed quark is produced with energy E', with a step function spectrum

$$f_{e^{\mp}}^{(1)}(E') = \frac{1}{kM_X} \theta(kM_X - E') \,. \qquad (27)$$

This lepton then decays into a positron of energy E with a similar spectrum, for each E',

$$f_{e^+}(E; E') = \frac{1}{kE'} \theta(kE' - E) \qquad (28)$$

with $k = 0.7$ as before. The resulting positron spectrum is the convolution of both of these distribution functions[21]

$$f_{e^{\mp}}^{(2)}(E) = (k^2 M_X)^{-1} \int \frac{dE'}{E'} \theta(kM_X - E')\theta(kE' - E) = (k^2 M_X)^{-1} \ell n\left(\frac{k^2 M_X}{E}\right). \qquad (29)$$

This is of course an approximation, and will break down for $E \leq 1$ GeV, below which the spectrum should flatten to a constant.

The second generation positron source function is then given by

$$Q_{e^+}^{(2)}(E) = n_x^2 <\sigma_x v>_A [\ 0.18\ \frac{1}{31} + 0.13\ (\frac{3}{31} + 2 \times \frac{27}{31}\ )]\ f_{e^+}^{(2)}(E)\ . \tag{30}$$

Expression (30) can be used in conjunction with eq. (19) to evaluate the second generation prompt positron spectrum numerically. The result, designated (P2), is shown in Fig. 2.

C. The Positron Flux from Charged Pion Decay.

The positron flux from the decay of π^+ mesons produced in heavy fermion annihilations can be determined by using the e^+e^- collider data[22]. Within experimental error, the data for the pion production spectrum from e^+e^- annihilations with cms energies above 14 GeV can be fitted to a single spectral function having the form

$$(s/\beta)(d\sigma/dx) \simeq 60 e^{-17x} + 7 e^{-6.9x} \quad \mu b\ GeV^2\ ,\quad x \equiv E_\pi/M_x . \tag{31}$$

The pion yield function $\zeta_\pi f(E_\pi) \equiv \sigma_{had}^{-1}(d\sigma/dE_\pi)$ for $M_x = 15$ GeV is obtained by dividing eq. (31) by the hadronic yield cross section $\sigma_{had} s = 4\pi\alpha^2 R/3 \simeq 0.35\ \mu b\ GeV^2$ to obtain

$$\zeta_\pi f(E_\pi) = \beta_\pi (11.6\ e^{-1.13 E_\pi} + 1.35\ e^{-0.46 E_\pi})\quad GeV^{-1}. \tag{32}$$

where, in the notation adopted here, ζ_π is the pion multiplicity per annihilation. The positron spectrum may be obtained from eq. (32) by noting that $E_{e^+} \simeq (1/4)E_\pi$. These positrons are soft enough so that energy losses by synchrotron radiation and Compton interactions may be neglected. This flux is given to a good approximation by

$$I_{(\pi)}(E) = \frac{Q_{(\pi)}(E)c\tau}{4\pi} = \frac{n_X^2 <\sigma_X v>_A c\tau}{4\pi} (5.4e^{-1.84E}). \qquad (33)$$

and is plotted in Fig. 2, labled (π). Fig. 2 also shows the total positron flux from Majorana fermion (generic $\tilde{h}$ or ν_M) annihilation. Fig 3 shows this interstellar flux together with the range of modulated fluxes obtained between minimum and maximum modulation as it varies over the solar cycle. There is little low energy positron data and future measurements will be obtained over some, as yet unknown, time during the solar cycle. The calculated positron flux is comparable to that calculated for cosmic-ray secondaries in the leaky box model[23].

The data for energies in the 1-20 GeV range[24], do not indicate a cutoff. Those results, however, may not be as accurate as one might hope, owing to the small number of detections involved and the difficulty in subtracting the atmospheric background. The measurements above 10 GeV energy are based on the detection of very few positrons and therefore have large error bars and are, in some cases, contradictory. Clearly, the experimental situation should be improved and clarified with new measurements, and eventually with an orbiting magnetic spectrometer experiment such as "Astromag"[25].

D. The Annihilation Neutrino Spectrum.

The cosmic ray neutrino spectrum from $\chi\chi$ annihilation can be obtained in a similar way to the positron spectrum, but with the following differences: (1) There are no electromagnetic energy losses so that only the neutrino source spectrum need be used, with the expected flux given by

$$I(E_\nu) = Q(E_\nu)<\ell>/4\pi \qquad (34)$$

with $<\ell>$ being the mean path length of the annihilation region. However, the fact that the b quark decay channels, which dominate the higher energy (prompt) neutrino production, involve b quarks which carry off, on average, 80% of the energy[11] (see footnote on pg. 13), has been taken into account. (2) There is no solar wind modulation of neutrinos, which also simplifies the problem. (3) One obtains the same number of ν_e's ($\bar\nu_e$'s) as positrons. This is also the case for ν_μ's ($\bar\nu_\mu$'s) from first-generation prompt (P1) decays, but the number of muon neutrinos from P2 and π decays is double that of the positrons. In addition, there is a component of muon neutrinos from kaon decay which has a harder spectrum than the pion-decay component so that it gives a significant contribution to the neutrino spectrum at about 2 GeV. Again, using the e^+e^- collider data[9], the kaon-decay neutrino spectrum as a function of ν_μ energy can be approximated by the expression

$$I_{(K)}(E) \simeq \frac{n_X^2 <\sigma v>_A c\tau}{4\pi} (0.73 e^{-0.76E}) \qquad (35)$$

The energies of all of the light leptons produced in each of the other processes respectively are similar. The low energy pion-decay neutrinos, which peak at ~35 MeV, can be calculated using the well-known kinematical formulae[26].

The resulting differential and integral spectra of ν_e's ($\bar\nu_e$'s) and

ν_μ's ($\bar\nu_\mu$'s) are shown in Figs. 4 and 5. The spectra are normalized to the upper limit on the γ-ray flux from the galactic center as given by Blitz, et al.[27] (see next section) assuming that γ-rays at this level and neutrinos could be from $\chi\chi$ annihilation. These fluxes may be, in fact, too pessimistic for two reasons: (1) the upper limits obtained by Blitz, et al.[27] may be too low by a factor of five or more[28], and (2) there may be a "hidden" dark matter annihilation neutrino source at the galactic center which could be opaque to γ-rays but not neutrinos. The flux of annihilation neutrinos can be compared with other predicted extraterrestial and atmospheric neutrino fluxes[29].

The differential ν_e production spectrum has been independently calculated using a Lund model[30] Monte Carlo program with the results given in Fig. 6 (Stecker and Tylka, in preparation). These results give a spectral shape which agrees well with the spectrum obtained from the analytic calculations given in Fig. 4.

The spectrum of annihilation neutrinos in the sun, which has been discussed by many authors[31], will be significantly different from the one shown in Figs. 4 and 5 owing to the fact that muons and π's interact in the solar interior before they decay. This eliminates the (π) and most of the (P2) components from the spectrum.

IV. ANNIHILATION GAMMA RADIATION

The spectrum of γ-ray background radiation from $\chi\chi$ annihilation in the halo may be calculated by noting that the continuum flux is overwhelmingly due to the decay of neutral pions produced in the $\chi\chi$ annihilations. One can then make use of the pion production spectrum (33) in order to determine the γ-ray spectrum.

The γ-ray spectrum resulting from the decay of the neutral pions is given by[21]

$$\zeta_\gamma f(E_\gamma) = 2\int_{E_\ell(E_\gamma)}^{M_\chi} dE_\pi (E_\pi^2 - m_\pi^2)^{-1/2} \zeta_\pi f(E_\pi) \tag{36}$$

where $E_\ell(E_\gamma) = E_\gamma + m_\pi^2/4E_\gamma$ and ζ_γ is the γ-ray multiplicity.

The resulting high latitude galactic γ-ray spectrum expected from $\chi\bar{\chi}$ annihilation is then

$$I(E_\gamma) = \frac{\langle \ell \rangle}{4\pi} n_\chi^2 \langle \sigma v \rangle_A \zeta_\gamma f(E_\gamma) . \tag{37}$$

The integral flux $I(>E_\gamma)$ is shown in Fig. 7 for two cases: (a) a mean line-of-sight integration length $\langle \ell \rangle$ of 10 kpc for the galactic halo for purposes of estimating the γ-ray background flux using the mean value of n_χ appropriate to the other calculations and (b) an isothermal halo model density of the form $4n_\chi^2(10 \text{ kpc})^2/(r^2 + a^2)$ evaluated looking perpendicular to the galactic plane at $r = 10$ kpc for a core radius $a = 10$ kpc. The production rate is again normalized to the cosmic-ray antiproton data. This eliminates the factor $(4\pi)^{-1} n_\chi^2 \langle \sigma v \rangle_A$. This integral flux plotted in Figure 7 is compared with the observations of the isotropic background and high galactic latitude galactic disk background[32] and an estimate of the background expected from bremsstrahlung and pion production by cosmic rays interacting with gas at high galactic latitudes[33]. The γ-ray flux spectrum has an absolute cutoff at $E_\gamma = M_\chi = 15$ GeV as shown in Fig. 8 so that a composite γ-ray spectrum should exhibit a "step" at this energy if the annihilation flux is at all comparable to the cosmic-ray secondary flux. (Given the discussion of the galactic ρ_χ distribution in Ref. 16, even the annihilation fluxes shown in Fig. 7 are most probably too optimistic, although still within the astrophysical uncertainties.)

The extragalactic and cosmological γ-ray background spectrum from neutral heavy fermion annihilation can also be calculated following the methods given by Stecker[34] and can be shown to be negligible compared to the observed extragalactic background shown in Fig. 7.

Fig. 8 shows the calculated values of the γ-ray production yield functions $\zeta_\gamma f(>E_\gamma)$ and $\eta_\gamma(>E_\gamma) \equiv (\sigma v)_A \zeta_\gamma f(>E_\gamma)$. A cosmic γ-ray source at a distance r_s consisting of dark matter fermions in a volume V_s with mean-square density $<n_\chi^2>$ will produce a flux

$$F(>E_\gamma) = (4\pi r_s^2)^{-1} <n_\chi^2> V_s \eta(>E_\gamma) \text{ cm}^{-2} \text{s}^{-1}. \tag{38}$$

The value shown in Fig. 8 for $\eta(>0.3 \text{ GeV})$ together with eq. (38) and the parameters given in the isothermal core model of Ipser and Sikivie[35], implies a γ-ray flux of $1.7 \pm 0.6 \times 10^{-7}$ cm^{-2}s^{-1} above 0.3 GeV from dark matter annihilating at the galactic center. This can be compared with the upper limit of 4×10^{-7} cm^{-2}s^{-1} derived in Ref. 17, indicating that dark matter annihilations may not contribute significantly to the γ-ray flux at the galactic center (See Ref. 28 for a detailed calculation of γ-ray spectra from the annihilation of various dark matter candidates having a range of masses from 5 to 20 GeV.)

The annihilation process $\chi\chi \rightarrow$ quarkonium + γ may produce potentially observable high energy γ-ray lines[36], but they will be intrinsically weak[37], so that searching for them will require a detector with very high energy resolution.

VIII. CONCLUSION

The annihilation of weakly interacting neutral fermions produces secondary cosmic-ray spectra with characteristic features, including a kinematic cutoff at energy $\sim M_X$. If such particles make up the dark matter in the galactic halo, study of the antiproton, neutrino, positron and γ-ray components of the cosmic radiation can provide potential tests for the particlular nature of the galactic dark matter. The antiprotons from the annihilation of heavy dark matter fermions have a production spectrum which peaks near 1 GeV energy and can produce a significant flux of low energy 150-300 MeV antiprotons after modulation by the solar wind. If ~15 GeV generic higgsinos or Majorana neutrinos make up the dark matter in the galactic halo, the entire reported spectrum of cosmic-ray antiprotons can be explained as the product of dark matter annihilation. Photinos may have too small a predicted annihilation cross section to easily give the required yields of cosmic-ray antiprotons, but higgsinos and heavy Majorana neutrinos can. A high energy cutoff at ~15 GeV is hinted at in the observed antiproton spectrum, but not firmly established. However, in any case, our calculations predict a cutoff energy below ~20 GeV, since weakly interacting fermions above this mass will not be abundant enough relics of the early universe to produce both heavy galactic halos and significant antiproton fluxes.

The cosmic-ray neutrinos, positrons and γ-rays from heavy fermion annihilation will also have a continuous spectrum with the same characteristic cutoff energy. The higher energy annihilation leptons come from prompt heavy quark and tau lepton decay processes, whereas the lower energy ones come from the decay of charged pions. The positron flux in the 1-10 GeV range is expected to be comparable to the observed flux and also comparable to that expected from the decay of charged pions produced in cosmic-ray interactions. It may be possible to look for a "step" in the total positron spectrum at ~15 GeV, given sufficient energy resolution. The same consideration holds for annihilation neutrinios and γ-rays. Pion decay γ-rays, have a spectral peak at ~70 MeV energy and pion decay neutrinos peak at ~35 MeV. These annihilation products are even more plagued by competition with cosmic-ray secondary production. Future balloon flights and space missions such as Gamma Ray Observatory and the proposed Astromag experiment should greatly increase the quality of the cosmic-ray antiproton, positron and γ-ray data. With future neutrino detectors, observations of annihilation neutrinos from dark matter fermions at the galactic center may be possible, although solar annihilation neutrinos should be seen more easily. It is to be hoped that future empirical studies will be able to settle definitively the question of the nature of the dark matter.

I would like to thank David Cline and the organizers of this workshop for inviting me and allowing me to share in such an earthshaking event.*

---

*During the meeting, a magnitude 6.1 earthquake occurred in the Los Angeles area on Oct. 1, 1987.

References

1. G. R. Blumenthal, et al., Nature **311**, 517 (1984); V. Trimble, Ann. Rev. Astron. Astrophys. **25**, 425 (1987).
2. H. E. Haber and G. L. Kane, Phys. Rpts. **117**, 75 (1985).
3. F. W. Stecker, et al., Phys. Rev. Lett. **55**, 2622 (1985).
4. S. Rudaz and F. W. Stecker, Astrophys. J. **325**, in press (1988).
5. H. Baer, et al., Phys. Lett **B183** 220 (1987).
6. E. W. Kolb and K. A. Olive, K. A., Phys. Rev. **D33**, 1202; erratum: Phys. Rev. **D34**, 2531 (1986).
7. M. Davis and J. Hucra, Astrophys. J. **254**, 437 (1982).
8. Q. Shafi and F. W. Stecker, Phys. Rev. Lett. **53**, 1292 (1984).
9. W. Bartel et al., Phys. Lett. **84B**, 444 (1980); R. Brandelik, et al., Phys. Lett. **104B**, 325 (1980); S. L. Wu, Phys. Rpts. **107**, 59 (1984).
10. D. Luckey, in A.I.P. Conf. Proc. No. 113, Experimental Meson Spectroscopy, ed. S. J. Lindenbaum, 271 (1984).
11. H. Aihara, et al., Phys. Lett. **B134**, 199 (1987).
12. A. Buffington, and S. M. Schindler, Astrophys. J. (Lett.), **247**, L105 (1981).
13. McDonald, F. B., et al., Proc. 19th Intl. Cosmic Ray Conf. **5**, 193 (1985).
14. J. S. Perko, Astron. and Astrophys., **184**, 119 (1987).
15. R. L. Golden, et al., Astrophys. Lett. **24**, 75 (1984).
16. K.-W. Ng, et al., U. Minn. preprint UMN-TH-589/86, Phys. Lett. **B**, in press.
17. J. Binney, et al., Mon. Not. Royal Astr. Soc. **226**, 149 (1987).
18. V. L. Ginzburg, and V. S. Ptuskin, Rev. Mod. Phys. **48**, 161 (1976).
19. F. J. Gilman, and S. H. Rhie, Phys. Rev. **D31**, 1066 (1985).

20. F. W. Stecker, Astrophys. and Space Sci. **3**, 579 (1969).
21. F. W. Stecker, Cosmic Gamma Rays, Mono Book Co., Baltimore (1971).
22. H. J. Behrend, et al., Z. Phys. C **20**, 207 (1983); W. Braunschweig, et al., Z. Phys. C, **33**, 13 (1986).
23. R. J. Protheroe, Astrophys. J. **254**, 391 (1982).
24. A. Buffington, et al., Astrophys. J. **199**, 669 (1975); J. L. Faneslow, et al., Astrophys. J. **158**, 771 (1969); R. L. Golden, et. al., submitted for publication in Astron. and Astrophys. (1986).
25. J. F. Ormes, et al., Proc. Workshop on Cosmic Ray Exper. for the Space Station pg. 124 (1985).
26. G. T. Zatsepin and V. A. Kuz'min, Sov. Phys. J.E.T.P. **14**, 1294 (1962).
27. L. Blitz, et al., Astron. Astrophys. **143**, 267 (1985).
28. F. W. Stecker, Phys. Lett B, in press (1987).
29. F. W. Stecker, Astrophys. J. **228**, 919 (1979).
30. T. Sjöstrand, Comp. Phys. Comm. **27**, 243 (1982).
31. K.-W. Ng, et al., Phys. Lett B **188**, 138 (1987) and references therein; S. Ritz and D. Seckel, preprint.
32. C. E. Fichtel, et al., Astrophys. J. **217**, L9 (1977).
33. F. W. Stecker, in The Large Scale Characteristics of the Galaxy, ed. W. B. Burton, p. 475, Reidel. Pub. Co., Dordrecht (1979).
34. F. W. Stecker, Astrophys. J. **223**, 1032 (1978).
35. J. R. Ipser and P. Sikivie, Phys. Rev. **D35**, 3695 (1987).
36. M. Srednicki, et al., Phys.. Rev. Lett. **56**, 263 (1986).
37. S. Rudaz, Phys. Rev. Lett. **56**, 2128 (1986).
38. E. A. Bogomolov, et al., Proc. 17th Intl. Cos. Ray Conf., **9**, 146 (1981).
39. E. A. Bogomolov, et al, Proc. 20th Intl. Cosmic Ray Conf. **2**, 72 (1987).

Figure Captions

Figure 1. The interstellar (extra-solar system) cosmic-ray antiproton flux from interstellar dark matter fermion annihilation (dashed line) and that spectrum modulated by the solar wind as discussed in the text (solid line) compared with the observed fluxes as measured by Buffington, et al.[12], (Bu), Bogomolov, et al.,[38] (Bo) and Golden, et al.,[15] (G). The lower curves, marked CRS, show the predicted flux of antiprotons as cosmic-ray secondaries produced by cosmic-ray collisions in interstellar space[23]. In addition to the data shown, there is a recent report of additional data of Bogomolov, et al.[39] in the 0.2-2 GeV range, (which must be compared to a curve modulated for a different part of the solar cycle) which appear to be consistent with the annihilation interpretation.

Figure 2. The cosmic-ray positron spectra from prompt first generation (P1) prompt second generation (P2) and pion decay positrons (π) produced following galactic annihilations of 15 GeV fermions. The flux is shown for the value $\kappa = 1.45$, chosen to fit the antiproton data. The total spectrum and flux is also shown along with the data (see text). The open squares are from Faneslow, et al.[24], the x's are from Buffington, et al.[24], and the solid circles are from Golden, et al.[24].

Figure 3. The total positron spectrum from Fig. 2 shown along with a range of modulation estimates between the maximum and minimum of the solar cycle using the method of Perko[14]. The data are also shown.

Figure 4. The differential neutrino flux spectrum from the galactic center normalized to the γ-ray upper limit as discussed in the text.

Figure 5. The integral neutrino flux spectrum from the galactic center normalized to the γ-ray upper limit as discussed in the text.

Figure 6. The differential electron-neutrino spectrum from the annihilation of generic higginos of mass 15 GeV obtained using the Lund Monte Carlo program (Stecker and Tylka, in preparation).

Figure 7. The high galactic latitude γ-ray background spectrum calculated for dark matter fermion annihilation in the galactic halo (see text) compared with the observed high galactic latitude isotropic and galactic "disk" radiation[31] and that expected from high galactic latitude cosmic-ray interactions[32].

Figure 8. The γ-ray yield spectrum per annihilation from the annihilation of 15 GeV higgsinos.

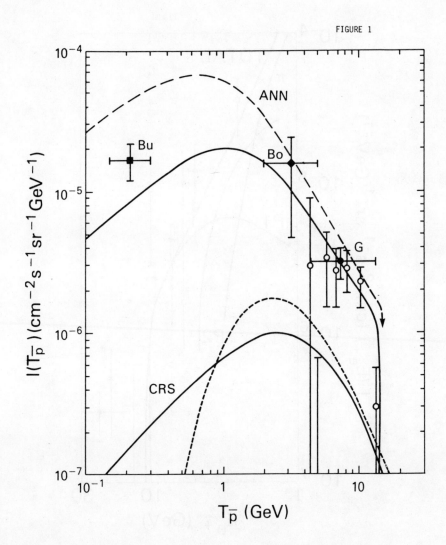

FIGURE 1

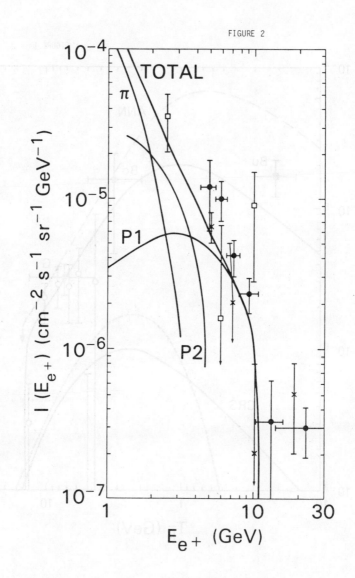

FIGURE 2

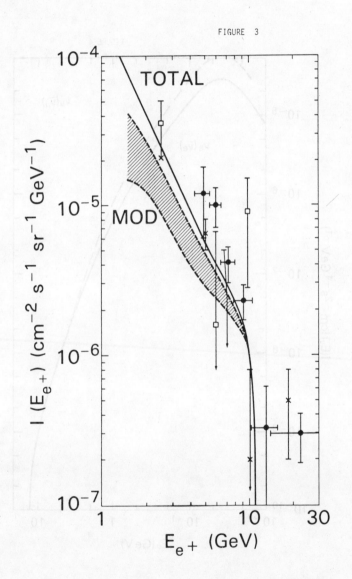

FIGURE 3

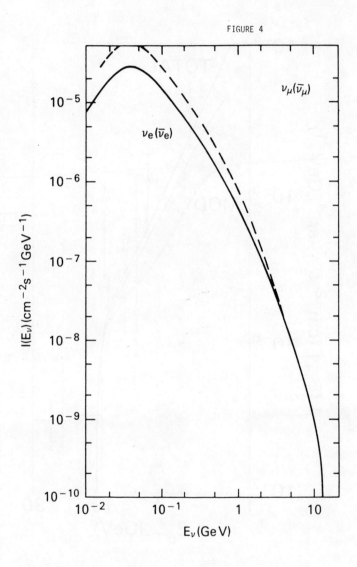

FIGURE 4

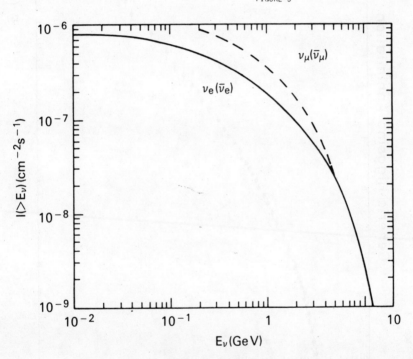

FIGURE 5

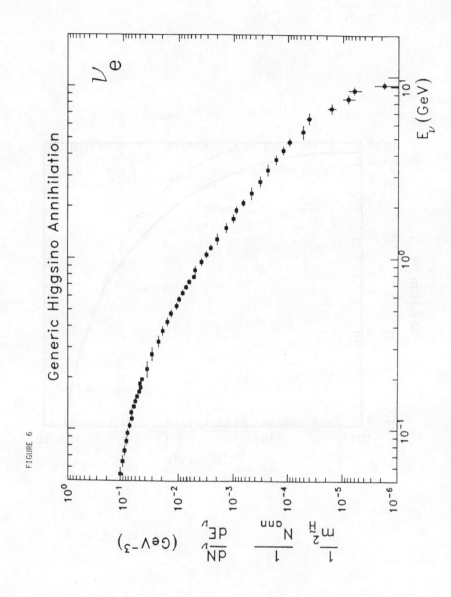

FIGURE 6

FIGURE 7

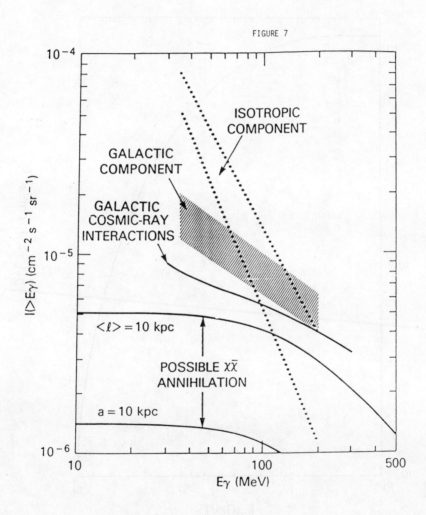

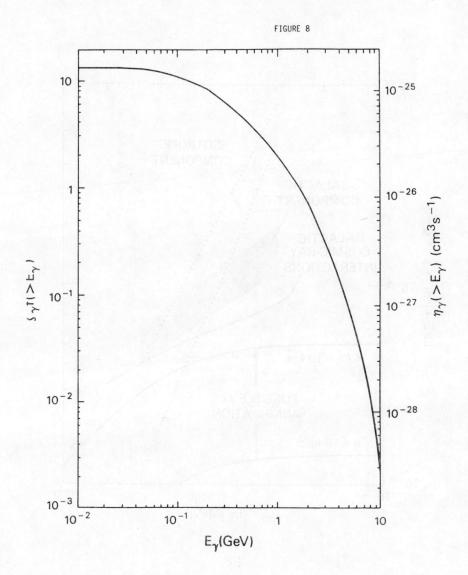
FIGURE 8

B. DETECTORS UNDER DEVELOPMENT.

Neutrino Pulse from Supernova
Inner Core: General Comments
on the Microphysics and Its Implications
for Pulse Duration and Spectrum

R.F. Sawyer

Department of Physics
University of California
Santa Barbara CA 93106

I. INTRODUCTION

If we added up all the things which it has been suggested can be learned or confirmed by studying the neutrino pulse from supernova 1987, we would end with a list with almost more entries than the number of events. Among the suggestions have been the following:

a) Confirmation of the hydrodynamical collapse model and a measurement of some of its parameters, such as the mass and temperature of the inner core.

b) Determination of the number of light neutrinos, and of their masses and stability.

c) Illumination of many aspects of the microphysics at high densities.

d) Determination of whether or not neutrinos have nonstandard interactions.

It is, of course, not possible to draw conclusions on every subject from a handful of data. In general, to learn anything, from the data, about a particular aspect of the process will require taking very seriously the results of a "standard" calculation of everything else. What's worse, even the most essential part of the theory, from the standpoint of analyzing the neutrino pulse, is not really in satisfactory condition; namely, the theory of how neutrinos get out of the inner core. There is probably no exact definition of "inner core", but collapse scenarios provide one that is appropriate for our considerations. It seems likely that roughly one-third of the neutrinos produced from electron capture on protons, within or without nuclei, escape within a few one-hundredths of a second, during collapse, having been produced in layers that are not yet sufficiently dense to trap the neutrinos. The other two-thirds of the lepton number is trapped (Arnett 1977) and diffuses out as electron neutrinos on a time scale variously estimated as between a few tenths of a second and a few seconds (Bethe, Brown, Applegate and Lattimer 1979; Sawyer and Soni 1979; Burrows and Lattimer 1985; Mayle, Wilson and Schramm 1987; Bludman, Lichtenstadt and Hayden 1982). The region of trapping is roughly that which will comprise the remnant, if all goes well and the outer layers are blown off, and we shall refer to it as the inner core.

The newly formed neutron star loses its larger excitation energy, by radiating neutrinos and antineutrinos of all small mass varieties, with a time scale which is in principle different from the time scale for diffusive deleptonization. The estimate given by Sawyer and Soni (1979) gives a time scale for cooling of roughly ten times that for core deleptonization. The more recent whole scenario calculations do not find as great a distinction. It could be an important distinction for design of future experiments, or even for analysing the results of SN1987A, and we shall come back to the question later in this talk.

Let me emphasize again the distinction which is being made. It is not between the "prompt" neutrinos from electron capture at densities lower than those producing neutrino opacity and the "trapped neutrinos" diffusing out of the opaque core at later times. It is rather between the diffusion of heat, by transport of all kinds of neutrinos and antineutrinos, and of lepton number, by diffusion of electron neutrinos. These phenomena are governed by somewhat different physics, and it is no surprise that the characteristic times are different.

In what follows, I shall discuss qualitatively aspects of the core physics, some not incorporated yet into the computational models for the whole supernova scenario; then I shall give some general conclusions for the time scales of lepton loss and cooling, for the spectrum of neutrinos, and for the connection to observations.

II. QUALITATIVE FEATURES OF THE NEUTRINO PHYSICS IN THE CORE WHICH AFFECT THE NEUTRINO PULSE

There are a number of qualitative features of the core physics which are worth pointing out:

a) Neutrons, protons, and electrons all will be degenerate in those core regions which are most important to the (1 sec) deleptonization pulse (the exact regions of degeneracy, in space and time, being dependent on detailed collapse scenarios).

b) Neutrinos will also be degenerate during the deleptonization era. Neutrino reaction rates are always reduced under such circumstances, from the denial of phase space; by the same token these rates increase rapidly with increasing temperature, as phase space is opened up.

c) The neutrino-electron interaction will not be a significant component of the opacities in the core regions.

d) Nuclei will have been disassociated throughout most of the inner core, during the early lepton loss and cooling era. However, depending very much on the results of detailed calculations, themselves depending on opacities, they may be present in the outer layers.

e) Exotic states of matter could be present in the higher density regions of the inner core. However, it seems reasonable to neglect this possiblity until the situation in which ordinary nuclear matter is assumed is better under control.

f) The interaction of neutrinos with nucleons in dense matter takes a number of different forms in the different domains of interest. The neutral current scattering of neutrinos off of neutrons and protons, or the coherent scattering off of nuclei, very nearly conserves neutrino energy. The equilibration of the ephemerally trapped neutrinos (to a distribution determined by the local temperature and lepton density) is accomplished most efficiently through neutrino absorption and emission processes; the one which comes to mind is $\nu + n \leftrightarrow e + p$. The usually neglected reactions,

$$\nu + n + n \leftrightarrow e^- + p + n$$

$$\nu + (\text{Nucleus}) \leftrightarrow e^- + (\text{Nucleus})'$$

may play a significant role as well. The latter, with the arrow going to the left, has been considered many times as a part of the collapse scenario, since the rates control the generation of entropy during infall. But its role in the outward diffusion of neutrinos has generally been ignored. As we shall see, the cross sections for neutrino absorption on hot nuclei become important in the transport problem, and new results will be presented. Furthermore, it will be the (equilibrating) neutrino absorption reactions which are most important in determination of the spectrum of neutrinos leaving the star, and in the definition of a photosphere.

g) Neutrino absorption cross sections in hot matter do not vanish for the case of vanishing neutrino energy; there is no window for the prompt escape of low energy neutrinos. An example is the three-body reaction, $\nu + n + n \to e + n + p$, (Sawyer and Soni 1979; Friman and Maxwell 1979; Haensel and Jerzak, to be published). In this reaction amounts of momenta limited only by the energies of the incident nucleons may be imparted to each of the final particles, even when the incident neutrino is arbitrarily soft. This process is likely to be quite important in the denser regions of neutron stars if the temperature is of the order of 10 MeV. The cross sections have been calculated by several authors; they do not vanish for $E_\nu = 0$. The extra nucleon is needed to transfer momentum in such a way that each participating fermion is within $k_B T$ of its Fermi surface. There is phase space of order $(k_B T)^2$ for the process, and since there are strong correlations in the nuclear wave function there is no problem in arranging for the momentum transfer.

Finally, let us consider the processes: $\nu + (\text{nucleus}) \leftrightarrow e^- + (\text{nucleus})'$ and $\nu + (\text{nucleus}) \leftrightarrow \nu + (\text{nucleus})'$ (Kolb and Mazurek 1979) where the incident and outgoing

64

nuclei can be ground or excited (or even dissociated) states. Just as in the previous case, the nuclear wave function can transfer the momentum necessary to make the reaction proceed.

III. NEUTRINO OPACITIES

Having made these qualitative remarks let me take a few minutes to go over some actual expressions for neutrino opacities arising from different mechanisms. The quantities of interest are certain reaction rates, τ^{-1}, for neutrinos of energy E_ν, in the medium. In the simplest case, neutrino absorption in a nondegenerate medium of nucleons, this rate would be given by the absorption cross section on a single nucleon times the density of nucleons. In other cases it will be defined with respect to the collision term in the Boltzmann equation for neutrinos in the medium.

For example, in the case of the absorption reaction $\nu + n \to e^- + p$ the rate, τ_a^{-1}, is just the differential absorption rate for a neutrino of energy E_ν, on a neutron, integrated over the appropriate occupancy factors for the three species. This reaction will only be kinematically allowed, under degenerate conditions, when the trapped lepton fraction exceeds a critical number, approximately $Y_L = .08$ for the case of matter near nuclear density. In this case a controlling parameter is the "driving" chemical potential

$$\mu = \mu_p + \mu_e - \mu_n \tag{3.1}$$

which will be the chemical potential of the electron neutrino, if there is complete beta equilibrium. We do not designate it as such, because the results which we shall quote on absorption rate are applicable to the surface layers of the body, as well as the deep interior, and near the surface there is significant deviation from beta equilibrium. The absorption rate, for a neutrino of energy E, in neutronized matter at near nuclear density, is given by (Sawyer and Soni 1979; Goodwin and Pethick 1982),

$$\tau_a^{-1} \approx 3.9 \times 10^5 \left[\pi^2 \left(\frac{kT}{1\text{MeV}}\right)^2\right](Y_L - .08)\theta(Y_L - .08) \times \left[1 + \exp(\mu - E_\nu)/kT\right]^{-1} \text{sec}^{-1} \tag{3.2}$$

But for the case of (neutral current) scattering of neutrinos from neutrons and protons, in the case of neutrino degeneracy, the situation is more complex; the transport equation no longer possesses the simple form involving only rates in the collision term (Sawyer and Soni 1979; Goodwin and Pethick (1982). Suffice to say, that in this case there still is a rate definable which is sufficient for estimates,

$$\tau_s^{-1} \approx 1.4 \times 10^5 \left[\pi^2 \left(\frac{kT}{1\text{MeV}} \right)^2 + \left(\frac{E_\nu - \mu}{1\text{MeV}} \right)^2 \right] \left(\frac{\mu_\nu}{100\text{MeV}} \right) \ sec^{-1} \quad (3.3)$$

Note the dependence on neutrino energy in both of the expressions (3.2) and (3.3).

Next we look at the neutral and charged current reactions for the case of no trapped leptons, $\mu = 0$ (corresponding to a $Y_L < 0.02$), and a neutron gas as the medium. For the neutral current scattering the answer is

$$\tau_s^{-1} \approx \frac{\sigma_0}{3\pi^2} c^{-4} (1 - c_A)^2 E_\nu^{\frac{3}{2}} \left[p_F^{(n)} \right]^2 \hbar^{-3} m_e^{-2} \left[1 + \frac{4(kT)M_N c}{p_F^{(n)} E_\nu} \right] \quad (3.4)$$

which leads to a numerical estimate (based on the second term only)

$$\tau_s^{-1} = 2.6 \times 10^{+3} \left(\frac{E_\nu}{1\,\text{MeV}} \right)^2 \left(\frac{kT}{1\,\text{MeV}} \right) \left(\frac{\rho}{\rho_{\text{Nuclear}}} \right)^{\frac{1}{3}} sec^{-1} \quad (3.5)$$

Most users have not included the temperature term on the right-hand side of (3.5); and it is easy to see that for thermal neutrinos this term is more important than the first. The absorption of a thermal neutrino on a neutron via the charged current interaction is exponentially suppressed under conditions such that e, p, n are degenerate; momentum cannot be conserved with each participating fermion lying within (kT) of its Fermi surface. This is where the three-body reaction, $\nu + n + n \rightarrow e + n + p$, comes in

$$\tau_3^{-1} \approx 1.8 \left(\frac{x^4}{4} + \frac{5}{2} \pi^2 x^2 + \frac{9}{4} \pi^4 \right) (1 + e^x)^{-1} \left(\frac{kT}{1\,\text{MeV}} \right)$$

where

$$x = (\mu - E_\nu)/k_B T \quad (3.6)$$

At temperatures greater than 15 MeV the opacity due to this process will exceed that due to neutral current scattering. The coefficient in (3.6) depends very strongly on the short range correlation function between nucleons. The above estimate is based on the conservative estimates of Friman and Maxwell. Note that even low energy neutrinos can be confined by this mechanism.

Finally, let me state a result for the absorption of neutrinos by nuclei. The model used for the nuclear transition strength is a generalization of that used by Bethe, Brown, Applegate, and Lattimer, whose model basically counts the states as in a Fermi gas, but concentrates the transition strength in a state (or a cluster of states) at a particular energy. The generalization is to the case of nonzero temperature and allows the transition strength to be distributed over a range in energies; this should be essential in regions of average density approaching nuclear density, regions in which the momentum transfer from the leptons to nucleons is of the order of 100 MeV. At a density of 10^{14} gm/cm^3 and a proton fraction of .3, we obtain, (in the Rosseland mean)

$$\lambda_a^{-1} \approx 1.9 \times 10^4 \left(\frac{T}{1\,\text{MeV}}\right)^2 \left(\frac{\mu_\nu}{1\,\text{MeV}}\right) \text{sec}^{-1} \qquad (3.7)$$

The opacities arising from (3.7) are large in comparison to those from all other processes in matter at the same densities, if a substantial fraction of the nucleons are in nuclei. Since current scenarios provide matter with large nuclear fractions in the outer regions of the inner core, the process will be important. Of note is the fact that the opacities are nonvanishing in the limit of zero neutrino energy.

IV. NEUTRINO TRANSPORT

Now, as I survey the literature on numerical neutrino transport calculations, I find more of the above results left out than included. I cannot jump to a conclusion as to what the effects would be if they were included. But since, I claim, it should be back to the drawing board one more time, for serious neutrino transport, let me take the opportunity to make the last presentation of a back-of-the-envelope calculation in this field.

We wish to estimate the time scales for deleptonization, by emission of electron neutrinos, and for cooling, by emission of all species of neutrinos and antineutrinos. Why are these time scales different? There are two reasons:

a) The opacities are different for electron neutrinos and for the other species, because of the absorption reactions discussed above.

b) More important, these time scales are not really those for a neutrino to "get out" by a process of bouncing off of scatterers in the medium. The quantities which are being transported are energy, on the one hand, and lepton number on the other. Although transported only by neutrinos, these quantities are not stored by the neutrinos alone: in the case of heat, the storage is almost entirely through the neutron specific heat;

in the case of lepton number, some of the storage is in electrons rather than in neutrinos.

To estimate the time scale for cooling, we begin from the expression for energy flux in terms of temperature gradient and (Rosseland) mean free time, $\bar{\tau}$,

$$\text{ENERGY FLUX} = \frac{7}{12} a c^2 T^3 \frac{\partial T}{\partial r} \bar{\tau}_s(T) \times N \tag{4.1}$$

where a is the black body constant and N is the number of species of light neutrinos. We estimate the temperature gradient as

$$\frac{\partial T}{\partial R} = T_{\text{central}} \times (\text{core radius})^{-1} \tag{4.2}$$

and use the value for mean scattering time which comes from the neutral current calculation, (3.5),

$$\bar{\tau}_s = \left(\frac{\rho_{\text{nuclear}}}{\rho}\right)^{\frac{1}{3}} \left(\frac{k_B T}{1\,\text{MeV}}\right)^{-3} \times 2 \times 10^{-5}\,\text{sec} \quad . \tag{4.3}$$

The reservoir of heat is taken to be,

$$\text{THERMAL ENERGY} = \int_0^{T_{av}} C_v(T) dT \times (\text{volume of core}) \tag{4.4}$$

where we take C_v to be the specific heat of a free neutron gas at nuclear densities,

$$C_v(T) = 10^2 \left(\frac{k_B T}{1\,\text{MeV}}\right) \text{erg/cm}^3 \,{}^\circ K \tag{4.5}$$

Taking an average temperature of 20 MeV, we estimate a cooling time,

$$\tau_{\text{cooling}} = (\text{stored energy})/(\text{rate of energy loss}) \approx 30\,\text{sec}\,/N \tag{4.6}$$

For the deleptonization process we follow the same steps, using the appropriate modification for lepton flux, rather than energy flux. We take, for $k_B T \ll \mu$,

$$\mathcal{F}_N = \text{neutrino number flux} = (6\pi^2)^{-2} \frac{\partial \mu}{\partial r} \bar{\tau}_a(T) \mu^2 \tag{4.7}$$

This equation defines the average rate $\bar{\tau}_a(T)$, which is computed from the Boltzmann equation with collision term derived from (3.2) (and the related expression for neutrino absorption; the two combine to eliminate the final thermal factor in (3.2)). The average rate turns out to be proportional to T^{-2}. Evaluated at nuclear density, and for $Y = 0.2$ we obtain

$$\bar{\tau} = 0.86 \times 10^{-7} \left(\frac{T}{1\,\text{MeV}}\right)^{-2}\,\text{sec} \tag{4.8}$$

The general formulae are in Sawyer and Soni (1979). The deleptonization time is estimated from the lepton number of the star, divided by (flux) × (area). Proceeding as before, taking a core radius of 20 kM, an average temperature of 10 MeV and a lepton number of 2.5×10^{56} we obtain,

$$\tau_{\text{deleptonization}} \approx 1\,\text{sec} \tag{4.9}$$

V. THE SPECTRUM OF EMITTED NEUTRINOS

The predicted spectrum of neutrinos in the pulse is of great interest in interpreting the events from S.N. 1987, and in the design of future experiments. Several of the groups who have performed calculations of the entire collapse scenario have given results for the energy spectrum, or even better, the energy spectrum as a function of time, for the emitted neutrinos. These are heroic calculations; it could seem ungenerous to state an opinion that the input neutrino microphysics is deficient, and in the end the changes we are recommending may make little difference in the results. In any case the discrepancies in various assumptions as to input physics will get adjudicated eventually, particularly in view of the new importance of the results. Whether or not the recent spectrum predictions turn out to be borne out in future calculations, it will be useful both for understanding the results of "complete scenario" calculations, and for interim estimates of neutrino spectra, to go back to calculations of neutrino transport, by itself, in fixed media. There are some general concepts which can be examined, most particularly that of a "photosphere", a fictitious surface from which neutrinos emanate with a thermal distribution. Is the expected distribution of the form characteristic of a photosphere? Some nearly analytically discussible models can give some insight into the extent to which it is.

The basis for what follows will be steady state transport, which not only demands that the medium hold still during the exit time for a neutrino, but that there is sufficient time to establish a nearly steady flow, changing only slowly, as the lepton excess, in the deleptonization era, or the temperature, in the early cooling era, decreases. This assumption, which is fairly solid for the case of the cooling era, lasting tens of seconds, is less so for the shorter deleptonization era. But it should be adequate for the qualitative estimates we shall make, and was implicit in the estimates of decay times made above.

Recall the calculation of radiative transfer of light in the stellar surface, at the level of idealization of the "grey atmosphere" of Chandrasekhar's classic book, *Radiative Transfer*. Suppose, to begin with, that the interaction of the photons with the matter is by their emission and absorption, and that the rates, which can depend on temperature, density, and composition, depend on energy only through the required thermal factor, $[\exp(-E/kT) - 1]^{-1}$. In this case the solution to the radiative transport problem yields a universal spectrum of photons leaving the surface, a spectrum which is very nearly the black body spectrum at an effective temperature which is determined from the outgoing flux. The flux itself is determined by the solution to an interior problem in which the transport equation can be treated in the diffusion approximation, with the spectrum calculation requiring solution of the full Boltzmann equation in the surface layer. In spite of the fact that the outgoing photons were produced at varying depths, in regions of greatly varying temperatures, the universal spectrum emerges, independently of composition, density and opacity gradients in the surface layer (as long as the absorption coefficient has only the minimally required energy dependence). The temperature distribution in the layer rearranges itself to make this the case.

Now, what can this have do to with the neutrino transport problem? The conventional wisdom might be that it has no relation at all, since neutrino cross sections are energy dependent above all else, and the universality depends on energy independence of cross sections. But here one should go more carefully; the literature has ignored the fact that for the cases of neutrino absorption in degenerate matter, and in hot nuclei, where the rates, (3.2) and (3.7) are dominated by the thermal terms, the collision term in the transport equation (after neutrino emission and absorption have been combined) is effectively independent of the energy of the neutrino energy.

Under this circumstance, the analogue of the Chandrasekhar calculation becomes of some importance: first let us consider the case $\mu = 0$ (no lepton excess over the 1-2% in normal neutron star matter, as in the cooling era following deleptonization). Writing a transport equation for the neutrino distribution function, in the presence of emission and absorption rates

$$\tau_a^{-1} = \tau_o^{-1}(1+e^{-x})^{-1} \quad , \quad \tau_{em} = \tau_0^{-1}(1+e^x)^{-1} \quad , \quad x = E_\nu/k_B T \qquad (5.1)$$

we obtain

$$\cos\theta \frac{\partial f(\vec{p},z)}{\partial z} = -[f(\vec{p},z) - f_e(|p|,T(z))]\tau_o^{-1} \qquad (5.2)$$

where f_e is the equilibrium Fermi function evaluated at the temperature of the medium, and θ is the angle between p and the vertical. (Note that it is not assumed that the neutrinos are in or near equilibrium; it is the interactions with the bath of e,n,p's that introduces the Fermi distribution function into (5.1).)

The rate constant τ_o can be an explicit function of depth as well as depend on the temperature, which is a function of depth. But in the steady flow situation, the temperature profile adjusts itself to make the flux the same at each depth. The solution proceeds as in the photon case; the result is a universal spectrum emerging from the surface, depending only on the flux; a spectrum which is different, but not very different from a Fermi distribution. The exiting neutrinos are not all emitted (i.e. have their last interaction) at a surface of the same temperature, or even nearly the same temperature. It is the energy independence of the absorption rate which creates an effective photosphere, with an effective temperature.

One can play the same game for the case of lepton transport, $\mu > 0$. In the general case we would have depth dependent functions $T(z)$ and $\mu(z)$. For a description of the deleptonization era we take an isothermal atmosphere for simplicity. Again the equation can be solved numerically, leading to a chemical potential profile (required to make the lepton flux the same at every depth in the surface layer), and to a spectrum of neutrinos leaving the surface. For a given temperature the distribution depends only on the flux out. There are more deviations from an equilibrium distribution in this case. Figures 1 and 2 show the calculated spectra and Fermi distributions with an effective chemical potential, chosen to give the same flux, for two choices of temperature.

REFERENCES

Arnett, W.D. 1977, *Ap. J.*, **218**, 815.

Bethe, H.A., Brown, G.E., Applegate, J. and Lattimer, J.M. 1979, *Nucl. Phys.*, A324, 487.

Bludman, S., Lichtenstadt, I. and Hayden, G. 1982, *Ap. J.*, **261**, 661.

Burrows, A. and Lattimer, J.M. 1985, *Ap. J.*, **307**, 178.

Chandrasekhar, S. 1960, *Radiative Transfer* (New York: Dover).

Goodwin, B.T. and Pethick, C. J. 1982, *Ap. J.*, **253**, 816.

Haensel, P. and Jersak, A.J. 1986, preprint, Copernicus Astronomical Center, Warsaw.

Kolb, E.W. and Mazurek, T.J. 1979, *Ap. J.*, **234**, 1085.

Mayle, R., Wilson, J.R. and Schramm, D.M. 1987, *Ap. J.*, **318**, 288.

Sawyer, R.F., Scalapino, D.J. and Soni, A. 1980, Neutrino '79.

Sawyer, R.F. and Soni, A. 1979, *Ap. J.*, **230**, 859.

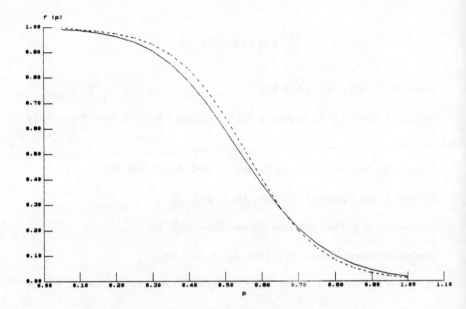

Figure 1: An effective occupancy factor, $f_e(p)$, is defined by

$$p^3 \frac{f_e(p)}{8\pi^2} = \frac{d\,\text{flux}}{dp} = \frac{p^3}{(2\pi)^2} \int_0^1 (d\cos\theta_p)(\cos\theta_p) f(\vec{p})$$

where the neutrino energy flux is the angular average appropriate to a spherical star, and $\cos\theta_p$ is the angle to the normal of the planar atmosphere.

In fig. 1, $f_e(p)$ is plotted for the case of a uniform temperature of 0.1 arbitrary units) and a flux which would correspond to a chemical potential, μ, of 0.56 (in the same units) if the distribution were thermal. The dashed line is the Fermi function for that thermal distribution. The units of p are those of T and μ.

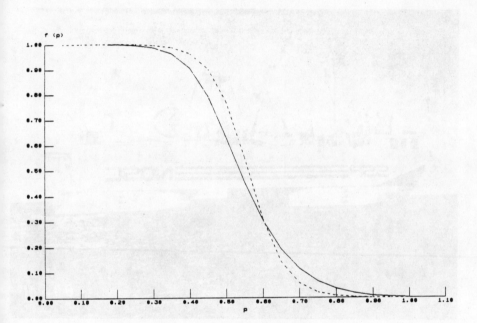

Figure 2: The same as fig. 1, except with $T = 0.05$.

DUMAND AND NEUTRINO ASTRONOMY

V.J. Stenger

University of Hawaii

ABSTRACT

The **DUMAND** project has pioneered in the progress toward neutrino astronomy. Best current knowledge implies that at least 10^4 m^2 detection area is required to detect neutrinos from the most promising sources: binary x-ray systems such as Cyg X-3. This is probably impractical underground, so the DUMAND concept of using the ocean as a detector of very high energy ($>$ 1 TeV) muon neutrinos remains viable. DUMAND has successfully completed its first stage with the deployment of a single string of phototube detectors and the measurement of cosmic ray muons at depths from 2000 to 4000 m. The basic detector units and technology required have been developed, ocean parameters and backgrounds measured and overall feasibility established. Proposals for the next stage are under study.

INTRODUCTION

Over 25 years ago, Markov[1] and Greisen[2] independently proposed that a large area detector placed underground, or underwater in a lake or the sea, would be capable of observing the muons produced by high energy neutrino interactions in the earth, and that these neutrinos could be of extraterrestrial origin. In order to achieve these aims, the Deep Undersea Muon and Neutrino Detector (DUMAND) was conceived in discussions at the Denver Cosmic Ray conference in 1973 and an informal group of interested people was assembled under the leadership of F. Reines, A. Roberts, S. Miyake and J. Learned. Shortly after, a series of workshops were held whose proceedings form perhaps the most complete set of early references in the field.

The first DUMAND Workshop, held in 1975 in Washington State, considered the concept of a large undersea array of photomultipler (PMT) tubes to detect the C̆erenkov light from the charged particles produced in neutrino interactions. A survey of possible sites led quickly to the conclusion that Hawaii is the best place for such an experiment, with deep, clear water near shore.

A two-week workshop in 1976 in Honolulu initiated the involvement of the University of Hawaii. This workshop gave considerable attention to the question of the detection of neutrinos from gravitational stellar collapse. It was determined that a 10^9 ton detector mass would be needed to observe supernovae at the rate of a few per year.[3,4] Even with the ocean as the detection medium, this appeared impractical. The more recent observation of neutrinos from SN1978a by much smaller underground detectors does not invalidate this conclusion; being the brightest supernova in 400 years, this apparently was a serendipitous event.

The 1976 workshop also saw the first extensive discussion of the possible cosmic sources of very high energy neutrinos (>1 TeV), including their emission from the expanding shell of a recent supernova.[5] The possibility of acoustic detection was also proposed at that time. A workshop in 1977 was devoted to this subject,[6] but acoustic detection was eventually shown to be impractical by DUMAND workers.

The proceedings of the DUMAND 1978 workshop in La Jolla contain, to my knowledge, the first references proposing neutron star binary systems, specifically Cyg X-3, as point sources of very high energy neutrinos.[7,8] It was recognized that there should be a connection with the Tev γ-rays which had been reported from a handful of sources.[9] At the same time, various diffuse source possibilities were examined and found to be unpromising.[10,11]

THE DUMAND FEASIBILITY STUDY

The great international interest in the DUMAND concept reached a peak in 1979 at a series of conferences in Japan and the USSR. This encouraged the University of Hawaii and the Department of Energy to support a Feasibility Study, begun in 1980. A symposium and two workshops were held that year which considered the scientific and technological problems in great detail. The results were published in four volumes of papers.[12,13]

As part of the Feasibility Study, a series of ocean-going experiments was initiated to test the detection concepts and measure the relevant oceanographic parameters of potential DUMAND sites. This was greatly aided by the strong oceanographic capability of the University of Hawaii Institute of Geophysics. A preliminary survey conducted in 1978 had acoustically mapped two prospective sites in Hawaiian waters, obtained bottom core samples, and measured the deep ocean currents.[14] Both sites were found to be adequate, with the one off the west coast of the island of Hawaii preferred. Water clarity measurements carried out in 1980 showed that the water at this site is exceptionally clear in the blue water wavelength region important for Čerenkov light: attenuation length = 28 m at λ = 450 nm.[15,16]

Unfortunately, we also learned the hard way what our ocean experts had tried to tell us: the ocean is a hostile place. On March 7, 1982, we lost an instrument, the *Muon String*, as its support cable broke in heavy seas. Still we persevered, and in the summer of 1983 successfully deployed an instrument to measure the bioluminescence background light level.[17] A significant depth relationship was discovered and higher light levels were observed while the instrument was being raised (Fig. 1). We interpreted this as the stimulated bioluminescence seen in the instruments's wake by the down-viewing phototubes.

This interesting result spurred us on to see how the bioluminescence would appear to PMT's sitting quietly on the bottom, as they would in any permanent array. In January 1984 the apparatus was deployed on the ocean bottom. Two independent timed explosive releases, each with supposedly better than 90%

probability of successful operation, failed to release the object from the bottom. We had lost our second instrument in less than two years.

Fortunately, DUMAND collaborators from the Institute for Cosmic Ray Research (ICRR) in Tokyo had been developing an independent experiment for measuring light background in the ocean. In summer, 1984 this instrument was deployed in a series of wonderfully successful operations at the DUMAND site.[18] First, the earlier results on stimulated bioluminescence were confirmed with ship-tethered measurements (Fig. 2). Then the instrument was placed on the bottom, with the PMT's 100 m off the sea floor, to observe quiescent conditions. This time the recovery was completely successful. The analyzed data showed that the light level 4.5 km deep, at the DUMAND site, was about an order of magnitude lower than observed at the same depth when the instrument was tethered to the ship. In fact, the measured light level of 218^{+200}_{-60} photons cm^{-2} s^{-1} was just slightly above what had been calculated and measured for the K^{40} in seawater, viz., 150 photons cm^{-2} s^{-1} for K^{40} (Fig. 2). Further, although an occasional bright pulse was seen, the light in general did not exhibit the time or pulse-height structure of the ship-tethered data (Fig. 3). We concluded that most of the biolight observed was stimulated, in disagreement with the predictions of experts. More important, for our purposes, was the conclusion that a bottom-moored array should not be washed out by bioluminescence.

This success for DUMAND was followed by another in summer, 1985, with the remarkable recovery of the instrument which had been lost at sea some 18 months earlier. The Scripps Institute for Oceanography vessel *Melville* had been cruising the Pacific picking up lost instruments. DUMAND personnel joined them in Hawaiian waters, aiding the crew locate and recover the instrument. Thus we were able to examine equipment which had rested for 18 months at the bottom of the ocean. No evidence was found for biofouling, and those parts of the instrument which had been properly treated prior to deployment were remarkably free of corrosion. Further, the data tapes were retrieved, analyzed, and found to give results on quiescent bioluminescence which were consistent with those found with the Japanese instrument. Thus we learned that the ocean is not so overpowering that we cannot fight back.

In 1982, the DUMAND collaboration had proposed to its various funding agencies that a staged program begin with the eventual purpose of placing a

500x250x250 m^3 array of 756 PMT's in the ocean for the primary purpose of very high energy neutrino astronomy (DUMAND Collaboration, 1982). Calculations indicated that if the γ-rays observed from several sources resulted from π^o production and decay, then a comparable flux of neutrinos above 1 TeV should be present (Fig. 4).[20] This implied that an array with an area of the order of 10^5 m^2 was needed to detect sources such as Cygnus X-3. This proposal was presented to the U.S. Department of Energy (DOE) in April 1983 who approved funding of the DOE-supported U.S. Groups for the first stage, the *Short Prototype String (SPS)*. NSF agreed to fund the Vanderbilt collaborators and ICRR in Japan was able to continue its significant role.

STAGE I: THE SHORT PROTOTYPE STRING (SPS)

The SPS is a string of seven optical detector modules and ancillary equipment deployed from a ship at variable depth.[21] As the first stage of DUMAND beyond the Feasibility Study, it had several purposes: (1) develop and test the basic detector module; (2) develop and test the associated technology, especially fiber optic signal communication; (3) learn more about backgrounds and other environmental parameters; (4) demonstrate that muons can be detected and their paths reconstructed by this technique; (5) measure muon depth vs. intensity with a homogenous overburden of matter. Only the last represented any attempt at obtaining new physics results. By the time of this conference, not all of this had been accomplished. Since then, a series of ocean operations in October and November, 1987, has produced excellent data at depths of 2000, 2500, 3000, 3500 and 4000 m (to be published). Thus Stage I has been brought to a successful conclusion.

The basic DUMAND detector unit, the optical module, is composed of a Hamamatsu 16-inch PMT encased in a glass pressure sphere. Electronic and power supplies are arrayed in two layers around the stem of the tube. The PMT output is converted to an optical signal whose leading edge specifies the time of arrival of the hit. The uncertainty in this time varies from 10 ns at the one photoelectron (pe) level to 5 ns for signals above 3 pe. The width of the optical output pulse specifies the collected PMT charge. The pulse width distributions are found to be Gaussian with $<PW> = 36 + 117\, q$ (ns) and $\sigma_{PW} = 15.4 + 36.2q$ (typically), where PW is the pulse width in ns and q is the number of photoelectrons. The angular response of the optical modules, including any blocking by the electronics, is is typically $0.55 + 0.45 \cos\alpha$, where α is the entry angle of a light ray.

The optical signal passes through the pressure sphere via a penetrator especially designed and manufactured by DUMAND personnel. Electrical power and 300 baud communications pass through a second penetrator. In addition to the signal processing circuitry in the optical module, sensors keep tabs on the environment inside the housing and a microprocessor monitors these parameter as well as providing experimenter remote control of the module circuits.

Operating at or near the single pe level, each optical module generates data at about a 100 KHz event rate from K^{40} background alone. Bioluminescence can produce even higher rates. The optical output of each detector module is carried by a multi-mode optical fiber to a central unit, the *String Bottom Controller (SBC)*, where it is multiplexed with others for transmission to the ship along a single mono-mode optical fiber. The SBC circuitry utilizes new gate array technology in order to handle the high data rates and ns timing of the signals. The fiber carrying these signals up to the ship is encased in a 6 km long 5/16-inch steel cable which also serves to support the string, send power from the ship with seawater return, and transmit 300 baud communications to and from the ship. This cable, designed for the U.S. Navy,[22] has other applications and represents another example of the development of useful new technology in the course of meeting the challenges of DUMAND.

The string also contains two modules which provide a calibrated light pulse for testing the optical modules are measuring the light attenuation length. An *Environmental Module* keeps track of depth, string orientation, acceleration and various environmental parameters including temperature and deep ocean current flow.[23] These data are transmitted to the ship via both an optical link and a much slower 300 baud link. The Environmental Module also transmits the outputs of two hydrophones which are mounted on the string to measure the acoustic background.

In order to minimize the stimulated bioluminescence as well as cable accelerations, we deploy the SPS from a stable platform, the U.S. Navy experimental research vessel *Kaimalino* (see frontispice). The Kaimalino is a

SWATH (Small Water Area Twin Hull) vessel which is exceptionally stable and thus able to operate in very rough seas.

VERY HIGH ENERGY NEUTRINO ASTRONOMY: THE BEST BET

The observation of *SN1987a* has certainly provided a welcome shot in the arm for neutrino astronomy. However, it must be recognized that this event was probably a stroke of luck which cannot be counted on to be repeated too often. The best bet for a continuing program of neutrino astronomy remains that outlined in the original 1982 DUMAND proposal: the detection of muons from very high energy (> 1 TeV) ν_μ interactions with large area underground or undersea muon detectors.

Observations of TeV and PeV γ-rays from a number of sources, mainly binary x-ray systems, strongly suggest that hadronic processes are involved. This implies a flux of ν_μ's from pion decay comparable to that observed for γ-rays from these sources. The ν_e fluxes should be much less. The preference for ν_μ over ν_e is fortunate. The great range of the very high energy muons produced in ν_μ charged-current interactions transforms any underground or undersea muon detector into very high energy neutrino telescope, with the earth or water surrounding the detector greatly multiplying the telescope's effective volume. In Fig. 5, the muon energy spectrum which would be expected from an extraterrestrial neutrino source is shown and compared with the background from atmospheric pion decays in the cosmic rays.[24] An E_ν^{-2} neutrino differential spectrum at the source is assumed, cutting off above 10^{16} eV. We see that most of the events have muon energies above 1 TeV, a result of two constructive effects: both the neutrino cross section and muon range increase with energy. The atmospheric neutrino background, on the other hand, has a steeper spectrum and the muons induced by these neutrinos are mostly below 1 TeV. The fact that there is no advantage, in fact a disadvantage, in being sensitive to muons below 100 GeV was important in the design of the proposed full-sized DUMAND array. Since virtually all of the muons will have ranges of hundreds of meters in water, the detector modules can be placed far apart, thus maximizing detector area with the minimum number of basic detector units.

In the case of Cyg X-3, the source most studied, calculations indicate that the ν_μ flux should be about three times the observed γ-ray flux above 1 TeV, mplying that a detector area of $\geq 10^4$ m^2 is needed for a signal of ten events per ear.[25] The largest existing underground detector, IMB, has an area of 400 m^2. he largest planned underground experiment, **MACRO**,[26] has a surface area of 400 m^2. These represent about the most one can practically achieve underground mines or tunnels. Thus the original DUMAND scheme of using a natural body f water as a Čerenkov detector remains the most viable to achieve the necessary etector size.

HE NEXT STAGE OF DUMAND

The 10^5 m^2 array envisaged in the 1982 DUMAND proposal[19] represents a ajor undertaking. Proposals have been made for smaller arrays to be deployed n the ocean bottom as the second stage of DUMAND. A three-string, 21 PMT, ray, dubbed the **TRIAD**,[27] would have an effective area for neutrino astronomy about 3000 m^2, larger than any planned underground detector. However, as escribed above, areas greater than 10^4 m^2 appear to be required for a reasonable xpectation of a signal given our current state of knowledge. A larger array with 100 PMT modules appears necessary, if the aim is to look for neutrino sources this signal level. The optimum configuration for such an array is not yet etermined. A four-string array of 64 modules has been studied in some detail;[27] would have an effective area of about 7500 m^2. After the results from the PS are analyzed and fed back into the Monte Carlo, these studies will be refined d a proposal for the next stage will be made. Since even 10^4 m^2 is still mewhat marginal, a 10^5 m^2 array remains the ultimate goal of DUMAND.

One possibility for a small-scale next step is to deploy a single string, rhaps the existing SPS, on the bottom. This would be relatively inexpensive, t still require the purchase of a 40 km long undersea electro-optic cable if the ployment is to reach 4.5 km. The possibility of deploying at shallower depths s been considered, but discarded because of the large cosmic ray background. ven at 4.5 km, a full 7-fold coincidence is required to limit the fake upcoming utrino events to below one per year, corresponding to an effective area of 200 ^{2}. It must also be noted that a single string provides only zenith angle formation, azimuth being undetermined.

An interesting added feature of a bottom-moored string, which might help stify its deployment, is the search for heavy particle candidates for the dark matter of the universe such as nuclearites (quark nuggets) and pyrgons (the particles associated with the compactification of higher dimensions). If these particles are of atomic dimensions and constitute the major mass of the galaxy they can emit sufficient light to be detected undersea with a few phototubes.[28]

COMPARISON WITH OTHER UNDERWATER PROPOSALS

Ideally one would like the clearest possible water for the detector. The ocean is unfortunately contaminated with salt and all natural bodies of water contain living organisms. As I have said, measurements at the proposed DUMAND site indicate that the background light under quiescent conditions near the bottom is essentially that produced by the radioactivity of the salt. Bioluminescence dominates only when the critters are disturbed.

The Soviets, who have carried on their own DUMAND project in parallel with the one reported here, report background light studies in the ocean consistent with those reported above.[29] They have also successfully deployed a string of PMT's at depths 850–1350 m in the fresh water of Lake Baikal, which has already set useful limits on magnetic monopoles.[30] Ultimately they plan to build a 4×10^5 m^2 area array and the project appears to be well-supported. In the U.S., a 62,500 m^2 water Cerenkov detector in a shallow lake or pit has been proposed.[31]

The Baikal experiment also reports that the attenuation length of the water in the lake is 20 m at 500 nm wavelength. Comparing this with 28 m for the water off Hawaii, we note that deep ocean sea water is clearer than fresh water in a natural lake, probably because of a higher density of biological matter in the lake.

Experiments in lakes or pits do not have the open hostility of the ocean to contend with, but pay a price for this convenience. Deployed in shallow water they must cope with the enormous cosmic ray muon background. Background reduction is accomplished by operating the phototubes at a high level of coincidence. In order to keep the coincidence time window sufficiently short, the tubes in coincidence must be close together. This clashes with the need to have tubes far apart to maximize the area. Thus the tradeoff is between going deep into the ocean, with the problems of ocean deployment, or building a shallow array in fresh water with perhaps an order of magnitude as many PMT's operated at

gh level of coincidence. At this time there is no obvious reason to decide which
the better option.

CONCLUSIONS

Although DUMAND has yet to see a neutrino, the DUMAND project has contributed significantly to the advance toward extrasolar neutrino astronomy. Much of the early work in neutrino source astrophysics and detector concepts was done in relation to DUMAND, with the proceedings of DUMAND workshops providing a major resource for the field. The notion of a deep ocean Čerenkov detector has been pursued by the international DUMAND collaboration, centered at the University of Hawaii. Our group of particle and cosmic ray physicists has gained considerable experience at operating in the ocean. We have recently succeeded in measuring muons in the ocean to a depth of 4000 m in what independent observers have called "the most sophisticated experiment ever done in the ocean." We have also made important measurements of ocean parameters and light backgrounds which will be invaluable to the development of any deep ocean installation and contributed to the development of a number of technologies which can be applied in the ocean and elsewhere.

The best bet for a continuing program of extraterrestrial neutrino astronomy remains the original DUMAND concept of detecting and reconstructing muons produced by the interaction of ν_μ's in the earth or sea. Unfortunately, ν_μ fluxes in the energy range where detection is optimum, above 1 TeV, are not expected to greatly exceed observed γ-ray fluxes, according to our best current estimates. In the case of Cyg X-3, an area $\sim 10^4$ m^2 is required, probably impractical anywhere on earth but underwater.

Experiments in comparatively shallow lakes and pits have been proposed which will have sufficient area, but require large numbers of phototubes operated at high coincidence levels to reduce the high background from downward cosmic ray muons. DUMAND would operate at a great depth in the ocean where the cosmic ray background is far lower, but must contend with a more hostile operational environment.

The DUMAND project has just completed its first stage, the Short Prototype String, which looked only at cosmic ray muons at great depths in the ocean. The

DUMAND approach to neutrino astronomy remains a viable alternative. If neutrino astronomy is worth doing, something like DUMAND will likely eventually be built.

ACKNOWLEDGEMENTS

DUMAND is a collaboration between the Universities of Bern, California, Irvine, Caltech, Hawaii, Kiel, Osaka, Purdue, Tokyo, Vanderbilt, Wisconsin. Many scientists, engineers and technicians at these institutions and others have contributed over the period covered in this review. Important contributions have also been made by personnel of the Hawaii Institute for Geophysics, Scripps Institution of Oceanography, and the Naval Oceans Systems Center. Special appreciation must also be given to the crew of the Kaimalino.

REFERENCES

1. Markov, M.A., Proc. 1960 Int. Conf. on High Energy Physics, Rochester 578 (1960)
2. Greisen, K., Ann. Rev. Nucl. Sci. 10, 63 (1960).
3. Tammann, G.A., Proc. 1976 DUMAND Workshop, Honolulu, 137 (1976).
4. Wheeler, J.C., ibid., 163.
5. Berezinsky, V.S., ibid., 229.
6. Bradner, H. (ed), Proceedings of the La Jolla Workshop on Acoustic Detection of Neutrinos (1977).
7. Eichler, D. and Schramm, D., Proc. 1978 DUMAND Workshop (Honolulu) 2, 135 (1978).
8. Helfand, D.J., ibid., 193.
9. Weekes, T.C., ibid., 313.
10. Silberberg, R., Shapiro, M.M., and Stecker, F.W., ibid., 231.
11. Stecker, F.W., ibid., 267.
12. Stenger, V.J. (ed), Proceedings of the 1980 International DUMAND Symposium (1980).
13. Roberts, A. (ed), Proceedings of the 1980 DUMAND Signal Processing Workshop (1980).
14. Harvey, R.R., Andrews, J.E., and Zaneveld, J.R., University of Hawaii Institute for Geophysics Report HIG-78-2 (1978).

15. Zaneveld, J.R., Proceedings of the 1980 International DUMAND Symposium, **1**, 1 (1980).
16. Bradner, H., and Blackinton. G., Applied Optics **23**, 1009 (1984).
17. Bradner, H. *et al.*, Hawaii DUMAND Center Report HDC-7-84 (1984).
18. Aoki, T. *et al.*, Il Nuovo Cimento **9**, 642 (1986).
19. DUMAND Proposal, Hawaii DUMAND Center (1982).
20. Stenger, V.J., Astrophys. J. **284**, 810–826 (1984).
21. Roberts, A. *et al.*, Proceedings of the DUMAND 1984 Workshop on Ocean Engineering and Deployment, 15 (1984).
22. Wilkins, *ibid.*, 135.
23. Elliott, J., *ibid.*, 147.
24. Stenger, V.J., Il Nuovo Cimento **9C**, 479 (1986).
25. Gaisser, T.K. and Stanev, T., Phys. Rev. Lett. **54**, 2265 (1985).
26. MACRO Collaboration, Il Nuovo Cimento **9C**, 281 (1986).
27. Stenger, V.J., Proc. of the Japan–U.S. Seminar on Cosmic Ray Muon and Neutrino Physics/Astrophysics Using Deep Underground/Underwater Detectors, Tokyo, 337 (1986).
28. Stenger, V.J., Hawaii DUMAND Center Report HDC-3-87 (1987).
29. Abin, A.V., *et al.*, Proceedings of the 20th International Cosmic Ray Conference, **6**, 273 (1987).
30. Bezrukov, L.B., *et al.*, *ibid.*, 292.
31. Gajeweski, W., *et al.*, University of California, Irvine, Proposal UCI 87-4 (1987).

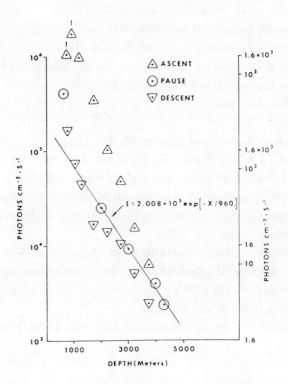

Fig. 1. Bioluminescent light intensity as a function of depth measured in the 1983 DUMAND ship-tethered experiment. The photomultiplier tubes look down and the greater intensity observed during ascent is interpreted as the stimulation effect of the instrument's wake. From Reference 17.

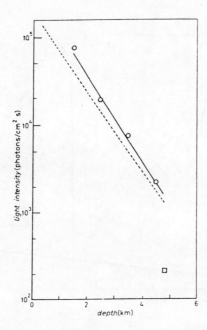

Fig. 2. Biomuminescent light intensity as a function of depth measured in the 1984 experiment. The circles are ship-tethered data. The square is the data point taken under quiescent conditions, with the instrument on the bottom. The level expected from K^{40} is 150 photons cm^{-2} s^{-1}. The dashed line is the previous ship-tethered result of Bradner *et al.*[17] From Reference 18.

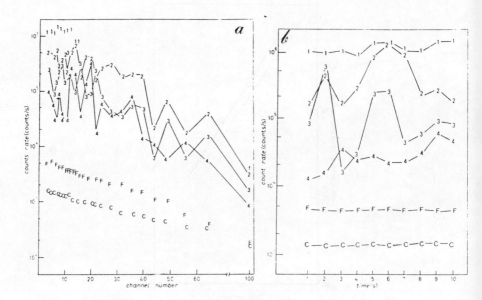

Fig. 3 (a) Pulse-height spectrum of background light from Reference 18. The data labeled 1-4 are for the depths 1500, 2500, 3500, and 4500 m respectively. The data labelled F are for the instrument moored near the bottom. The data labelled C are for dark noise measured at $3^{\circ}C$ in the laboratory; (b) Time variation in the count rate for these data.

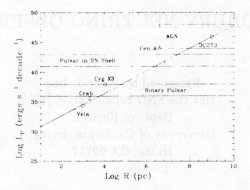

Fig. 4. The range of proton energy luminosities for types of possible very high energy neutrino sources and some specific examples. The diagonal line shows the detectability level for a full-scale DUMAND array. From Reference 20.

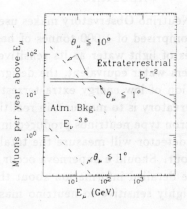

Fig. 5 The solid muon spectrum (solid) which would be measured by an underground or undersea muon detector from a source, such as a binary pulsar, which has an E_ν^{-2} differential neutrino spectrum. The flux is normalized to give 20 events per year. Two cuts on the radius θ_μ of the muon circle on the celestial sphere are shown. The dashed curve shows the background from ν_μ's produced in the atmosphere which would be observed in a detector of area 1000 m^2, in each of the two muon circles. From Reference 24.

THE SUDBURY NEUTRINO OBSERVATORY

Presented by Peter J. Doe
(for the SNO collaboration [1])
Dept. of Physics,
University of California, Irvine,
Irvine, CA 92717

ABSTRACT

The proposed Sudbury Neutrino Observatory makes use of an imaging water Čerenkov detector comprised of 1,000 tonnes of heavy water and approximately 1,800 tonnes of light water as its sensitive volume. Located at a depth of 5,900 meters water equivalent, the design of the detector is optimised for the detection of low energy extra terrestial neutrinos. The primary goal of the observatory is to measure, in real time, the flux, spectra and direction of electron type neutrinos produced in the Sun by the ^8B cycle. In addittion the detector will measure the total flux of all neutrinos, independent of flavour. Should a supernova occur within our galaxy, the detector will provide detailed information about the dynamics of the stella collapse and be highly sensitive to neutrino masses and oscillation parameters.

1 INTRODUCTION

The radiochemical experiment of Davis et. al. [2] has shown that there is a discrepancy of at least a factor of three between the experimentaly observed and the theoreticaly predicted flux of ^8B neutrinos produced in the Sun. This result is further supported by the recent data from the Kamioka II detector [3]. Further more, the latest radiochemical results suggest a possible anti-correlation between the ^8B neutrino flux and 11 year period of the Solar sunspot activity. These results are extremely difficult to accomodate within the standard solar model(SSM) [4]. Explainations of these experimental results fall into two catagories; new, though somewhat unsatisfactory, models of the solar dynamics [5] and new and exciting properties of neutrinos such as neutrino oscillations [6]. To distinguish between these diverse explainations of the Solar Neutrino Problem (SNP), it is clear that more detailed, real time information is required.

The detection of neutrinos from SN1987A by the IMB and Kamioka II water Čerenkov detectors [7] provided confirmation of the general theory of stella collapse, established new limits on the properties of neutrinos as well as demonstrating the power of such detectors. What is required is more detailed information on the spectra and time structure of all neutrino types produced in stella collapse, in order to further guide theory on this fundamental phenomenon.

The Sudbury Neutrino Observatory (SNO) will address both the above questions. Its primary goal is to measure in real time and with high statistics, via two independant reactions, the direction, spectral shape and flux of the ^8B neutrinos above a threshold of 5 MeV. In addittion a third reaction is avaliable to measure the total flux of neutrinos, independent of flavour, above a threshold of 2.2 MeV. Should a supernove occur within our galaxy, the unique capabilities of the SNO detector to distinguish between neutrino types would provide detailed information on the dynamics of the collapse and enable stringent limits to be set on the masses of the different neutrino flavours.

2 THE DETECTOR

The design of the SNO detector is driven by the need to reduce radioactive backgrounds in order to achieve its physics goals. Located at a depth of 2,000 meters (5,900 mwe) in the Creighton Mine, near Sudbury, Ontario, the cosmic ray muon flux will be approximately 100 per day through the central sensitive volume of the detector. Thus, such problems as muon spallation backgrounds will not be present. A schematic outline of the detector is given in Figure 1.

The cavity in which the laboratory will be built is a domed cylinder,

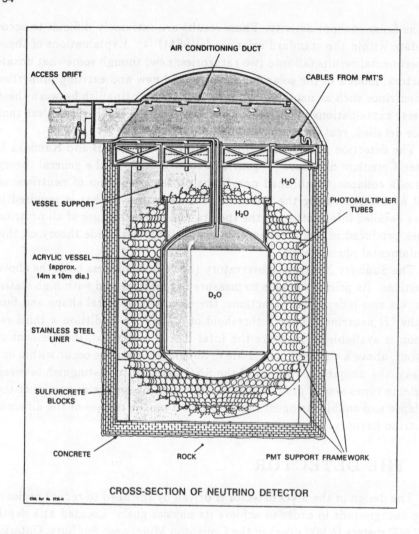

Figure 1: Schematic View of the SNO detector

20 meters diameter, by 32 meters high. In order to attenuate the gamma ray backgrounds associated with the surrounding rock, the cavity is lined with one meter of low activity, boron loaded, sulfurcrete. This in turn supports a stainless steel tank containing 5,000 tonnes of light water (H_2O), which further attenuates external radioactivity. The levels of uranium and thorium contaminants in the H_2O will be reduced to the level of 1 part in 10^{-14} by means of resin beds and manganese coated acrylic fibers. At a distance of 1 meter from the steel tank are located 2,000, 50 cm diameter photomultiplier tubes (PMT), providing 40% photocathode coverage. These PMT's collect the Čerenkov light resulting from charged particles interacting within the detector. At a distance of 2.5 meters from the PMT's is located an acrylic vessel which houses 1,000 tonnes of 99.8% enriched heavy water (D_2O). The distance of 2.5 meters is required to attenuate gamma rays originating from the PMT's and to allow accurate spatial reconstruction of the Čerenkov light source. The acrylic vessel is a 10 meter diameter right cylinder with hemispherical end caps, suspended from the deck of the cavity. This vessel must allow efficient transmission of the UV Čerenkov light, while contributing minimaly to the radioactive backgrounds and providing maximum security for the D_2O. To exclude the mine environment (in particular dust and radon gas) a system of air locks is incorporated, terminating in the gas tight deck covering the detector. This careful design results in the central volume of the detector being probably the lowest radioactive environment in the world.

The resolution of the detector is criticaly dependent on the performance of the PMT's. With a PMT timing uncertainty of 7ns (FWHM) our Monte Carlo studies indicate a spatial and angular resolution of 70cm and 25^0 respectively. With 40% photocathode coverage, an energy resolution of 20% at 7 MeV is predicted. Our ability to identify and reject backgrounds depends strongly on these parameters. The natural radioactive backgrounds also determine the trigger threshold, which at the present radioactivity levels is expected to be 5 MeV.

Our measurements indicate that at present, the uranium and thorium in the acrylic vessel (measured to be a few parts in 10^{12}) are the principle sources of background to the study of solar neutrinos.

3 THE REACTIONS

To address the SNP, three independent reactions are avaliable, making SNO a uniquely powerful device;
The neutrino-electron elastic scattering (ES) reaction,

$$\nu_x e \longrightarrow \nu_x e$$

is available to all tracking detectors containing electrons, and is the reaction of the Kamioka II detector. It offers excellent directional information, the scattered electron being kinematically constrained to a 30^0 forward cone. Although all neutrinos participate in this reaction, ν_e dominate by a factor of approximately six due to their higher cross section. In addittion, the recoil electron may carry any energy up to the maximum energy of the incident neutrino, thus it is difficult to extract from this reaction spectral information concerning the incident neutrino.

The charged current (CC) reaction,

$$\nu_e + d \longrightarrow p + p + e^-$$

which has a cross section is approximately an order of magnitude higher than the ES reaction, offers excellent spectral information with potentialy good statistics. Furthermore, only electron type neutrinos (which are expected from the ^8B reaction) may participate. Essentially all the neutrino energy, above the Q value of 1.44 MeV is carried by the electron, which is emmitted with an angular distribution of

$$1 - \tfrac{1}{3}cos(\theta)$$

with respect to the neutrino momentum. Thus, coarse angular information is also avaliable from this reaction.

Finaly, the neutral current (NC) reaction,

$$\nu_x + d \longrightarrow n + p$$

in which all neutrinos participate equally, will be used to measure the total flux of neutrinos above the threshold of 2.223 MeV (the binding energy of the deuteron). The reaction is observed via the energetic electrons produced by the gamma rays resulting from the capture of the thermal neutron. Due to the relatively low capture cross section of neutrons on deutrium and also the low energy (6.25 MeV) of the capture gamma ray, the detector has a low efficiency (20%) for this reaction. The situation may be inproved by the addition of chlorine in the form of salt (NaCl) which has a high capture cross section and produces 8.6 Mev of gamma energy. Since any gamma ray above 2.2 MeV may disintigrate the deutron, this reaction is extremely sensitive to radioactive backgrounds.

In the case of a supernova, two additional reactions are avaliable to measure the $\bar{\nu}_e$ flux,

$$\bar{\nu}_e + d \longrightarrow n + n + e^+$$

which takes place in the D_2O and has a Q value of 4.03 MeV, and,

$$\bar{\nu}_e + p \longrightarrow n + e^+$$

which takes place in the H_2O and has a Q value of 1.8 MeV.

4 SOLAR NEUTRINOS

A popular explaination of the SNP is that of matter enhanced neutrino oscillations as proposed by Mikheyev, Smirnov and Wolfenstein (refered to as the MSW effect) [8]. This offers a wide range of neutrino mass differrences and mixing as solutions to the SNP. Also, for many solutions, the spectral shape of the ^8B neutrinos is significantly distorted, offering further evidence as to the nature of the SNP. Thus, to distinguish between these and other solutions, it is particularly important to measure the spectral shape of the ^8B neutrinos.

The data taking scenario for SNO consists of three separate phases; first a H_2O fill which will provide only ES events, followed by a D_2O fill, providing ES, CC and, with reduced efficiency, NC events. Finally, to enhance the efficiency for detecting the NC events, NaCl will be added to the D_2O. This scenario follows a somewhat natural order and provides a maximum of information in a timely fashion. Using a detailed Monte Carlo of the Čerenkov detector and the radioactive backgrounds we have studied the performance of the detector under this data taking scenario, the results are summarised below.

Due to the great depth, low background and trigger threshold, the detector is seen to be highly sensitive, using only the H_2O. Since there will be no NC or CC events, this data will provide the ES signal (which will be strongly peaked in the forward direction) plus the CC backgrounds (which will be uniformly distributed) to be subsiquently used in conjunction with the CC signal. The angular distribution and spectral shape of the ES electron and the CC background, assuming a ν_e flux of $2\times10^6\nu_e$ cm/sec is shown in Figure 2. After one year of running, as a result of our extreamly low backgrounds, we have a flux sensitivity of $5\times10^{-5}\nu_e$/cm/sec, or 8% of the SSM prediction. It is important to note, in the light of possible modulation of the solar flux, that the ES data offers a sensitive monitor of flux changes throughout the life of the experiment. Figure 3 shows the response of the ES signal to a flux modulation corresponding to a change from 1/3 to the full SSM flux over a period of 12 months.

The subsequent data sets, taken with D_2O and D_2O + NaCl, contain ES, CC, CC background, NC and NC background events. We separate these reactions by use of two dimensional fits in energy and spatial/angular distribution. From the H_2O data we already have the shapes associated with the ES and the CC background. The spectral shape and efficiency for the neutron capture associated with the NC events will be obtained very accurately by inserting neutron sources in the D_2O. It is not possible to distinguish between a neutron which is liberated by a neutrino interaction and one associated with a background event, therefore it is important that we measure the backgrounds to this process. External sources of background may be estimated from the spatial distribution of events and internal sources of background (due to Uranium and Thorium in the components of the detector) will be obtained by measuring the activity in samples of the detector materials. We estimate that we will measure these backgrounds to ±25%.

Figure 4 illustrates how the fitting proceedure yields the separate components out of the total data set obtained after one year of running with pure D2O and assuming a flux of $2\times10^6\nu_x$/cm/sec (1/3 SSM) for the ES and CC events and a flux of 6×10^6 (the full SSM) for the NC events. In Figure 5 is shown the spectral shape and angular distribution of CC events. Although no MSW process has been assumed, it can be seen from the error bars that the data is sensitive to distortions in the spectral shape. In the

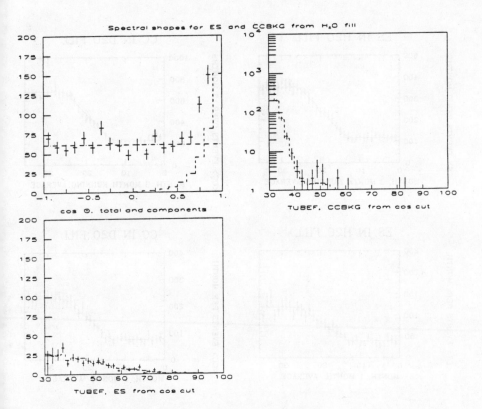

Figure 2: Elastic Scattering Data.
The angular distribition and spectral shape of the elastic scattering (ES) signal are shown on the left hand side. The spectral shape of the charged current background (CCBKG) is shown on the right hand side. TUBEF referes to the number of PMT's used in the event reconstruction and is a measure of the event energy.

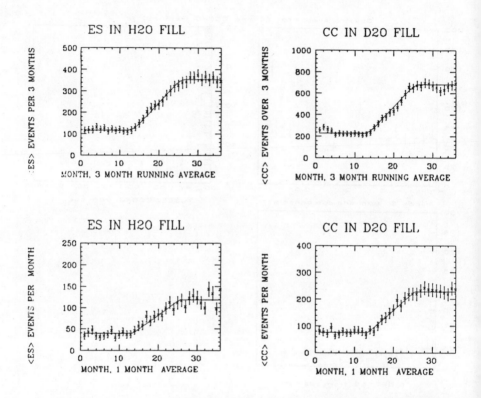

Figure 3: Flux monitoring.
The elastic scattering (ES) and charged current (CC) signals provide detailed information on flux variations. The solid line shows the change in the number of events starting with 1/3 SSM and increasing to the full SSM.

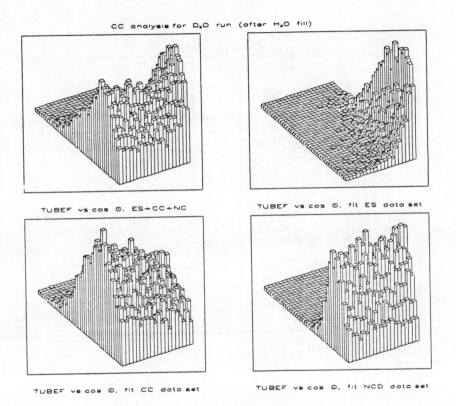

Figure 4: Separating the signals.
The figure shows the summed and individual signals in D_2O. The 3D view of elastic scattering (ES), charged current (CC) and neutron capture (NC), obtained from a 2D fit of a one year data set, in a plot of energy (number of PMT's used in fit, TUBEF) vs. direction ($\cos\theta$).

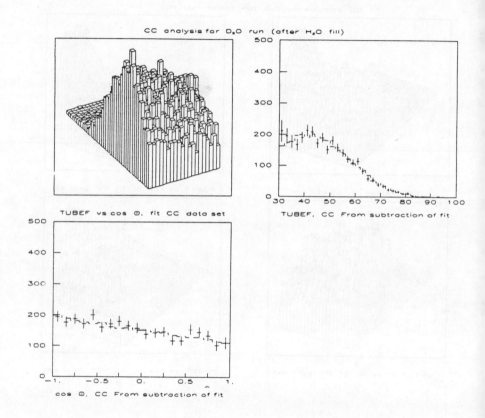

Figure 5: Charge current (CC) energy and angular distribution.

plest scenario for oscillations, one can assume that the ^8B neutrinos produced at the rate predicted by the SSM, but that in transit from Sun they have oscillated into equal parts of electron, muon and tau trinos. To address this question, one can examine the ratio of ES and events. Should there be no oscillations occuring and the flux of ^8B is ely produced at one third of the SSM, then the ratio of ES/CC rules the simple oscillation case by 16σ. Alternatively, should the oscilla- described above occur, then the ratio ES/CC differs by 4.8σ from the of no oscillations. In addition, the presence of the NC reaction, which sures the total flux of neutrinos (independent of oscillations) above 2.2 V is avaliable to further clarify the possibly complex situation arising 1 the possibility of solar neutrino oscillations.

The CC reaction may also be used as a monitor of the solar flux in the e manner as the ES reaction, the result is shown in Figure 3.

SUPER NOVA NEUTRINOS

Although not the primary goal of SNO, should a supernova occur whithin galaxy, the detector would provide unique data, based on its ability to inguish between the various flavours of neutrinos and its low trigger gy (a lower energy cut is possible for events occuring in a burst). The licted number of events in the SNO detector as a result of a stella col- e at 10 kpc is given in the table below. The sensitivity to ν_μ and ν_τ,

Reaction	Target medium	Events in ν_e burst kt^{-1}	Events in cooling phase kt^{-1}
$\nu_e + d \longrightarrow p + p + e^-$	D_2O	10	33
$\nu_x + e \longrightarrow \nu_x + e$	D_2O/H_2O	1	16
$\nu_x + d \longrightarrow \nu_x + p + n$	D_2O	6	760
$\bar{\nu}_e + d \longrightarrow n + n + e^+$	D_2O	0	20
$\bar{\nu}_e + p \longrightarrow n + e^+$	H_2O	0	120

nct

that accessable to terrestial experiments are possible.

6 CURRENT STATUS OF SNO

The SNO collaboration, formed in 1984, to take advantage of the possibility of obtaining on loan 1,000 tonnes of D_2O and the presence of a deep location for the laboratory, consists of institutes from the US, Canada and the UK. Much of our effort has been spent in simulating the response of the detector to neutrinos and radioactive backgrounds and using these results to refine the design of the laboratory and detector. The need for unprecedentedly low backgrounds has required pioneering work in locating low activity materials and radioactive measurement and purification techniques. Full engineering design studies have been made of the laboratory and the acrylic vessel. In addittion an exploratory drift into the region where the laboratory will be excavated, has been carried out, along with a geotechnical survey. The details of this work may be found in the Sudbury Neutrino Observatory Proposal [9].This proposal has been subject to an international review committee and a decision by the various agencies is pending.

References

[1] The SNO Collaboration consists of;
University of California Irvine: R.C. Allen, G. Buehler, P.J. Doe, F. Reines.
Princeton University: R.T. Kouzes, M.M. Lowry, A.B. McDonald.
University of Pennsylvania: E.W. Beier, W. Frati, F.M. Newcomer, R. Van Berg.
Queens University at Kingston: G.T. Ewan, H.C. Evans, J.R. Leslie, H.B. Mak, W. McLatchie, B.C. Robertson, P. Skensved.
National Research Council of Canada: J.D. Anglin, M. Bercovitch, W.F. Davidson, C.K. Hargrove, H. Mes, R.S. Storey.
Chalk River Nuclear Laboratories, AECL: E.D. Earle, G.M. Milton.
University of Guelph: P. Jagam, J.J. Simpson.

Laurentian University: E.D. Hallman, R.U. Haq.
Carlton University: A.L. Carter, D. Kessler.
Oxford University: D. Sinclair, N.W. Tanner.

[2] R. Davis Jr. et al., Phys. Rev. Lett. 20, 1205 (1968). The current status of this experiment was reported at Neutrino '88 (Boston, June, 1988) and will appear in the preceedings.

[3] See "Search for ^8B Solar Neutrinos at KAMIOKANDE-II" M. Nakahata, Ph.D. Thesis, University of Tokyo, Febuary, 1988.

[4] J.N. Bahcall et al., Rev. Mod. Phys. 54, 767 (1982), and also reference 5.

[5] For a current review of solar models see "Solar Models, Neutrino Experiments and Helioseismology", J.N. Bahcall and R.K. Ulrich, Astrophysics Preprint Series, IASSNS-AST 87/1. Submitted to Rev. Mod. Phys., September 1987.

[6] S.P. Rosen and J.M. Gelb, Phys. Rev. D34, 969 (1986).
W.C. Haxton, Phys. Rev. Lett. 57, 1271 (1986).
E.W. Kolb et al., Phys. Lett. 175B, 478 (1986).
V. Barger et al., Phys. Rev. D34, 980 (1986).

[7] K. Hirata et al., Phys. Rev. Lett. 58, 1490 (1987).
R.M. Bionta et al., Phys. Rev. Lett. 58, 1494 (1987).

[8] L. Wolfenstein, Phys. Rev. D17, 2369 (1978) and Phys. Rev. D20, 2634 (1979).
S.P. Mikheyev and A. Yu. Smirnov, Nuovo Cimento C9, 17 (1986).

[9] The Sudbury Neutrino Observatory Proposal, SNO-87-12, October 1987.

Neutrino Detection with MACRO at Gran Sasso

The MACRO Collaboration[†]

Presented by Charles W. Peck
California Institute of Technology
Pasadena, California 91125 U.S.A

1. Introduction

The MACRO detector, shown schematically in Fig. 1, is designed as a multipurpose deep underground high energy particle detector optimized for maximum sensitivity in the collection of penetrating high energy cosmic radiation. The present design consists of three thick horizontal scintillator planes with 4.5 m separation between planes; vertical scintillator planes covering the four sides; 18 layers of horizontal streamer tubes interspersed through the detector; five vertical streamer tube layers covering the scintillators on the detector sides; passive absorber to set the muon energy threshold and identify penetrating particles; and a layer of track etch detectors. MACRO modules are 12 m wide and will extend for approximately 72 m along Hall B of the Gran Sasso laboratory, which is located underground at a depth of 3600 m water equivalent. The principal goal in the design of the apparatus is that it be a sensitive and highly redundant detector for GUT monopoles. However, it is also an effective detector for high energy neutrinos from the atmosphere and from astrophysical point sources, and for the burst of low energy neutrinos from stellar collapse.

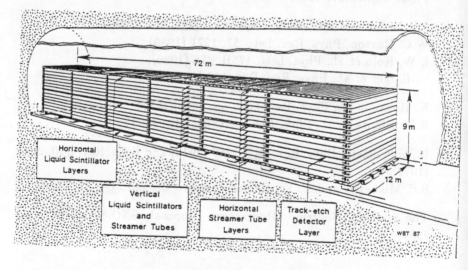

Fig. 1. Schematic Diagram of the MACRO Detector

2. High Energy Upward-going Neutrinos from the Atmosphere

In MACRO, neutrinos are detected by interactions with the rock below the detector which result in upward-going muons. These muons lose energy in the residual rock, and as a result, the collection volume increases with increasing energy. The combination of this, the increase of the ν-nucleon cross section with energy, and the hard spectral index of 2.0

(inferred from the γ-ray measurements), results in a most probable detected neutrino energy above 1 TeV.

MACRO's design gives the capability of adequately discriminating against the large flux of downward-going muons, so that no serious systematic problem is expected in measuring this upward-going muon flux. This leads to the possibility that these events could be used to search for neutrino oscillations by using the Wolfenstein-Mikheyev-Smirnov effect [L. Wolfenstein, *Phys. Rev.* **D17**, 2369 (1978) and S. P. Mikheyev and A. Yu. Smirnov, *Sov. J. Nucl. Phys.* **42**, 913 (1985)].

Following the method of Gaisser and Stanev [T. K. Gaisser and T. Stanev, *Phys. Rev.* **D30**, 985 (1984)] and assuming there are no matter oscillations, we have calculated the expected number per year of upward-going muons with energy greater than 2 GeV in the full MACRO apparatus. These are given in the following table, in which ψ is the nadir angle.

	$0 \leq \psi \leq 60°$	$60° \leq \psi \leq 90°$
μ^+	74	65
μ^-	150	132

Estimates [G. Auriemma *et al.*, to be published] of the changes in the expected rate as a function of the muon charge, momentum, and nadir angle have been made using a realistic model of the radial dependence of the Earth's density and various assumptions of the neutrino-antineutrino mass difference and mixing angle. The result is that after a five-year run, it would be *statistically* possible to exclude differences of squared mass greater than about 10^{-2} eV^2 if the mixing angle is greater than about 10^{-2}. However, this result depends upon the ability to *systematically* calculate the expected flux with no mixing to a precision better than what is probably reasonable. On the other hand, if MACRO were modified to allow charge separation for momenta less than 10 GeV, this dependence on calculated fluxes is significantly relaxed and the statistics are also significantly improved so that the experiment looks feasible; a five-year run would allow lowering the mass limit to about 10^{-3} eV^2. The required modifications to MACRO are being studied.

3. The Detection of Astrophysical Point Sources of Neutrinos

In recent years, multi-TeV gamma rays have been detected from a number of astrophysical sources such as Cygnus X3, Vela X1 and LMC X4. If it is true that the high energy γ-rays are a result of π^0 decay in hadronic cascades, then neutrinos must also be produced from the decay of the accompanying charged pions, and their flux can be estimated from the observed γ-ray fluxes. Except at very high energy, neutrinos will not be strongly attenuated in the companion, and so should be observable over the entire eclipse. Enhancement of the neutrino flux over the γ-ray flux can be as large as 30. This enhancement is assumed later when counting rates are given.

The features of the MACRO detector that should make it the first capable of observing neutrino point sources are: (1) its large surface area (3240 m^2), resulting in an adequate counting rate, (2) the superb angular resolution of its streamer tube system (~ 0.2 deg) which makes point sources stand out clearly above a uniform background, and (3) the ability of its scintillator system to separate several hundred upward-moving muons from $\sim 10^7$ downward-going muons per year through accurate time of flight measurements (~ 1 ns) over a flight path of over 9 m. The kinematics of muon production and the multiple scattering of a muon at the typical energies of ~ 1 TeV result in 90% of the muons being contained within a 1 deg cone. The background from atmospheric neutrinos in such a cone is about 0.1 count/year. On the basis of the observed γ-ray fluxes, we would expect between 5-10 counts/year from the two southern hemisphere sources, Vela X1 and LMC X4, well above the estimated background rate.

4. Detection of Gravitational Collapse within Our Galaxy

The large volume of liquid scintillator ($\sim$ 1000 m^3) contained in MACRO makes it possible to detect gravitational collapse events through the neutrinos that are emitted. Core neutronization results in about 10^{53} ergs of energy release in $\sim$10 MeV ν_e's over a time scale of milliseconds, whereas subsequent core cooling will result in about the same energy released in the form of $\sim$10 MeV $\nu\bar{\nu}$ pairs of all flavors emitted over a time scale of many seconds.

The ν_e's can be detected via neutral current interactions with electrons, while antineutrinos emitted during core cooling can be detected through their interaction with free protons in the liquid scintillator. In a typical gravitational collapse event located at the center of our Galaxy, we expect to see about 125 e^+ events in the 5-20 MeV region produced in MACRO over several seconds. Triggers are being prepared to detect any such event.

5. Outlook

Construction of the detectors has begun. The experimental hall at the Gran Sasso Laboratory is scheduled for completion by the fall of 1987. Installation of the first MACRO supermodule (12 m long × 12 m wide × 4.5 m high) will commence immediately thereafter and is expected to be completed by spring of 1988. Following testing and optimization studies, the remainder of MACRO will be completed over a two-year period.

CHARACTERISTICS OF THE "SMART" 35 CM DIAMETER PHOTOMULTIPLIER

D. Samm (*)

III. *Physikalisches Institut, RWTH Aachen*
Aachen, Fed. Rep. of Germany

Abstract

A new type of large photomultiplier has been developed. In this device photoelectrons from a hemispherical cathode are accelerated by 25 kV towards a scintillator layer which is optically coupled to a conventional photomultiplier. The resolution of the pulse height distribution for single photoelectrons is 58 % FWHM. The measurements show that it is possible to discriminate between one, two, three and more than three photoelectrons with high efficiency. The distribution of the transit time shows a resolution of 3.48 ns FWHM for single photoelectrons.

Introduction

In this paper some properties of a 35 cm diameter photomultiplier (PMT) are presented. This kind of phototubes is of great interest in the field of large scale underwater experiments, such as the search for the proton decay, observation of high energy neutrino point sources and of neutrinos from supernova or the detection of other deep water light events, like bioluminescence or radiation from radioactive nuclides.

The photomultiplier should have the capabilities of wide area detection for maximum response to faint (Cherenkov) light, single photoelectron detection with good energy resolution and fast timing. To fulfill these requirements tubes have been designed according to a new principle [1]: a high voltage is applied at the photocathode and the resulting high photoelectron energy is transformed into many photons which are detected by a conventional photomultiplier. In addition the high voltage will make the tube less sensitive to magnetic fields than traditional PMT's.

2. The Design of the Photomultiplier

The design of the large PMT ("preamplifier" tube) is shown in Fig. 1. It consists of a glass bulb of 35 cm diameter with a recess at the back and a hemispherical photocathode. The photocathode is of a bialkaline type SbKCs because of the high spectral sensitivity for blue light and the low thermionic emission. The photoelectrons (Pe) are accelerated by 25-30 kV towards a scintillation layer and the resulting light pulses from the scintillator are detected by a conventional fast photomultiplier, like the XP2972 placed in the recess. To minimize pulse pile-up at high count rates, the scintillator is chosen to be Yttrium silicate, Cerium activated, with a 1/e decay time constant of 35 ns. In this way one photoelectron from the large cathode will give rise to up to 30 photoelectrons in the small photomultiplier. The high voltage for the large PMT is placed at the back of the tube.

3. The PMT Characteristics

Several properties of a PMT are important, such as quantum and collection efficiency, gain, dark current, uniformity, energy resolutio and timing.

The quantum efficiency at 400 nm is 25 %. Due to the high accelerating voltage the collection efficiency is 100 %. The gain is measured by comparing the single photoelectron pulse height distributio of the small photomultiplier alone with the preamplifier/small PMT combination. It is a linear function of the high voltage, with a threshold voltage around 5 kV [2].

3.1 The Dark Current

The dark pulse counting rate should be low for a good signal/nois ratio. The integral dark current as a function of the high voltage is shown in Fig. 2. Up to about 30 kV it rises linearly with the high

ltage. In order to get the dark pulse counting rate one has to
nsider that many small pulses are generated (Fig. 3). The discrimi-
tion level of 1/2 photoelectron equivalent was set for the elimination
 these tiny dark pulses. Counting rates between 5 kcps and 50 kcps
e found at a temperature of 20°C. The dark pulse is mainly due to
ermal emission from the photocathode. Its dependence on the operating
mperature shows the expected behavior as can be seen from Fig. 4.

2 Uniformity

.When a large photomultiplier is designed, it is a natural question
ether a uniform sensitivity over the photocathode area is obtained or
t. In order to measure the uniformity the photocathode was illuminated
 a light spot through an optical fiber and the PMT was rotated to
ange the illuminated point on the photocathode around its spherical
nter. Figure 5 shows the relative uniformity as a function of polar
gle for two different azimuthal angles. The variation of uniformity
 to 90° is within 20 %.

3 Energy Resolution

In general, the energy resolution of a photomultiplier depends on
veral factors e.g. the number of input photons, the quantum efficiency,
e collection efficiency and the multiplication factor. To get the
ergy resolution the pulse height distributions were measured by using
blue LED with a pulse rate of 5000 Hz and with a sampling time of
0 s. The intensity of the LED pulses have been chosen, that the peaks
r two and three photoelectrons are visible (Fig. 6). From Fig. 7 it
n be seen that the single electron resolution improves with increasing
ltage in the large tube. At 30 kV the resolution of the single
ectron peak is about 56 % FWHM.

4 Timing Characteristics

One of the most important quantities determining the quality of a
otomultiplier is the transit timespread. Ideally the transit time of
otoelectrons should not depend on their point of origin on the photo-

cathode. The optimized design of the preamplifier tube as well as the high accelerating voltage minimize the transit time differences. The average transit time for the photoelectrons is only about 10 ns.

The timespread of the photomultiplier depends on the energy of the photoelectron, its emission angle and its emission point on the photocathode. Transit timespreads have been measured with the scheme shown in Fig. 8. The photocathode has been illuminated at several measuring points with light intensities yielding 1, 2, 5 and 14 photoelectrons on the average. In the worst case with 1 photoelectron on the average, the FWHM of the distribution is 3.48 ns. For 2, 5 and 14 photoelectrons the transit timespreads are 2.52 ns, 1.92 ns and 1.38 ns. Figure 9a, b shows the transit timespread distribution at one particular measuring point for average numbers of photoelectrons of 1 and 5 respectively. For the various measuring points the center of the distribution varies by at most ±0.7 ns.

4. Tests under Realistic Conditions

The first tests of the PMT's under realistic conditions have been performed in the Lake Baikal Experiment [3] during March-April 1987. A general view of the experimental set up is shown in Fig. 10 [4]. The underwater equipment consisted of a muon telescope and two "SMART" photomultipliers (8 and 10 as indicated in Fig. 10). The muon telescope detected vertical muons by coincidence technique. The Cherenkov light from these muons was detected by the PMT's. The muon telescope and the PMT's were attached to different strings with a spacing of 6.5 m between the strings. Figure 11 [4] shows the pulse height distribution of the muon induced Cherenkov light. The distribution shows that the most probable value of the number of photoelectrons is N = 5 which is in good agreement with the expected value of N = 4.9 [4].

5. Conclusion

In summary the large diameter phototube shows a very good single photoelectron resolution and excellent timing characteristics. The good energy resolution offers the possibility of using less tubes for the

etermination of the energy in particular for a low light signal such
s in widely spaced arrays for high energy neutrinos or in supernova or
olar neutrino detectors. Furthermore with the accurate timing it is
ossible to reconstruct the trajectory of a particle traversing the
edium and to reduce the background.

*) This work has been done in collaboration with P.C. Bosetti, III.
hysikalisches Institut, RWTH Aachen, Fed. Rep. of Germany, G. van Aller,
-O Flykt, W. Kühl and P. Linders, Phillips Electronics Components and
aterial Division, Eindhoven, The Netherlands and R. Kurz, K.D. Müller,
. Reinartz, A. Scholz and S. Widdau, Kernforschungsanlage Jülich, Fed.
ep. of Germany.

REFERENCES

1] G. Aller et al., IEEE Trans. Nucl. Sci, Ns-30 (1983) 469.
2] P.C. Bosetti, presented at the Workshop on Neutrino Masses and Neutrino Astrophysics, Ashland, 1987.
3] L. Bezrukov et al., presented at the 20th Int. Cosmic Ray Conference, Moscow, 1987.
4] L. Bezrukov et al., presented at the Workshop on Underground Experiments, Baksan, 1987.

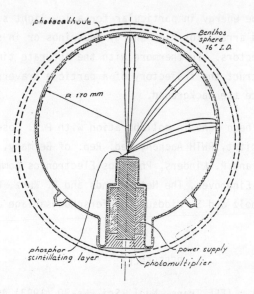

Fig. 1.   The schematic drawing of the 35 cm diameter PMT.

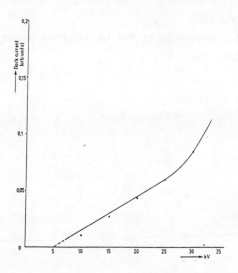

Fig. 2.   Dark current as a function of the accelerating voltage.

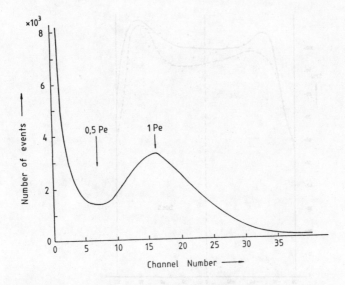

Fig. 3. Pulse height distribution of the dark pulses.

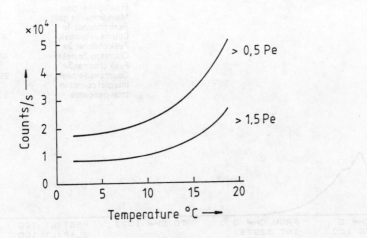

Fig. 4. Temperature dependence of the counting rates of dark pulses.

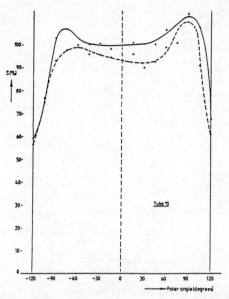

Fig. 5. Relative sensitivity as a function of polar angle, for two azimuthal angles.

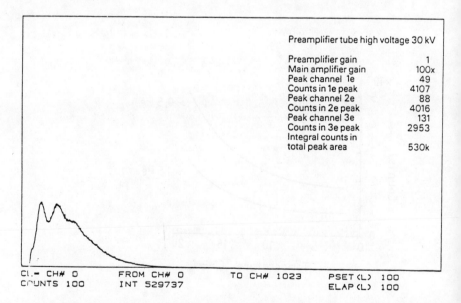

Fig. 6. Pulse height distribution for 30 kV.

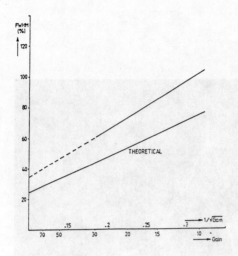

Fig. 7. FWHM resolution as a function of gain. Theoretical curve and measurements.

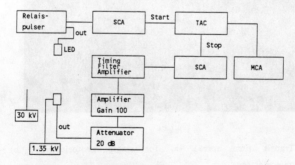

TAC : Time to amplitude converter
SCA : Single channel analyser
MCA : Multi channel analyser

Fig. 8. Measuring scheme of transit time spread.

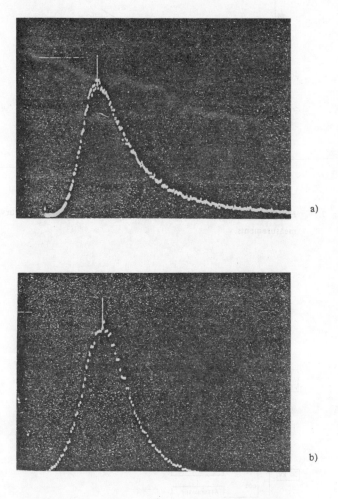

Fig. 9. Transit time spread, a) for single photoelectrons, b) for 5 photoelectrons.

119

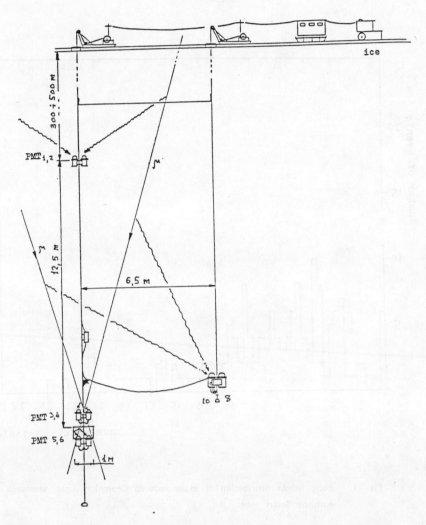

Fig. 10. A general view of the experimental set up of the Lake Baikal experiment.

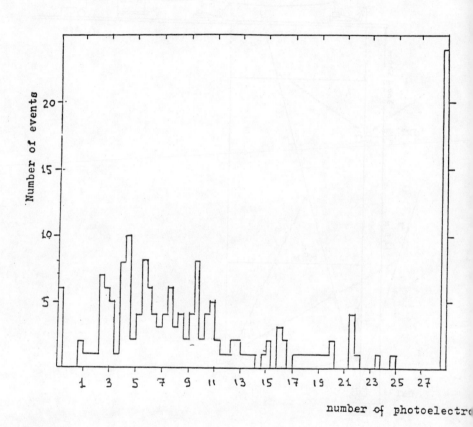

Fig. 11. Pulse height distribution of muon induced Cherenkov light, measured with the "Smart" tubes.

C. RESULTS FROM EXISTING DETECTORS

THE MONT BLANC DETECTION OF NEUTRINOS FROM SN 1987a

M.Aglietta[a] G.Badino[a] G.Bologna[a] C.Castagnoli[a] A.Castellina[a] V.L.Dadykin[b] W.Fulgione[a] P.Galeotti[a] F.F.Kalchukov[b] V.B.Kortchaguin[b] P.V.Kortchaguin[b] A.S.Malguin[b] V.G.Ryassny[b] O.G.Ryazhkaya[a] O.Saavedra[a] V.P.Talochkin[b] G.Trinchero[a] S.Vernetto[a] G.T.Zatsepin[b] V.F.Yakushev[b]

a) Istituto di Cosmogeofisica del CNR, Torino, Italy, and Istituto di Fisica Generale dell'Università di Torino, Italy
b) Institute of Nuclear Research, Academy of Sciences of USSR, Moscow, USSR

(Presented by P.Galeotti)

ABSTRACT

In this paper we discuss the event (5 interactions recorded during 7 seconds) detected in the Mont Blanc Underground Neutrino Observatory (UNO) on February 23, 1987, during the occurrence of supernova SN 1987a. The pulse amplitudes, the background imitation probability, and the energetics connected with the event are reported. It is also shown that some interactions recorded at the same time in other underground experiments, with a lower detection efficiency, are consistent with the Mont Blanc event.

1. INTRODUCTION

An Underground Neutrino Observatory (UNO) has been built[1] by our Institutes with the aim to search for bursts of low energy neutrinos from stellar collapses. The UNO is a Liquid Scintillation Detector (LSD), running[2] since October 1984 in the Mont Blanc Laboratory, at a depth of 5200 hg/cm^2 of standard rock underground. The very large coverage of rock, and an additional shield, provide very good background reduction, and allow us to operate the UNO at a very low energy threshold.
An event, considered as a candidate of a neutrino burst, was detected[3] on real time during its occurence, and identified before optical observations from the computer print-out on February 23.12

($2^h52^m36^s$ UT), 1987. Optical data indicate that no star brighter than magnitude 12 was present earlier than February 23.08, and that the supernova had a magnitude 6.1 at February 23.44.
Preliminary results of our detection have been published[4-6] previously; here we report the energy values of the 5 pulses in the burst, and discuss the background imitation probability of the event. We compare also our data with the results obtained in other underground experiments, and discuss the connected neutrino outflow from the stellar collapse which originated supernova SN 1987A. It is shown that the Mont Blanc event, besides being self consistent, it is also additionally supported by all the other experimental evidence available.

2. THE MONT BLANC UNDERGROUND NEUTRINO OBSERVATORY

The Mont Blanc neutrino telescope, located at the depth of 5,200 hg/cm^2 of standard rock underground in a cavity along the road tunnel linking Italy and France, is a Liquid Scintillation Detector (LSD) consisting of 72 counters (1.5 m^3 each) in 3 layers, arranged in a parallelepiped shape (see fig.1) with 6x7 m^2 area and 4.5 m height. The total active mass is 90 tons of the liquid scintillator

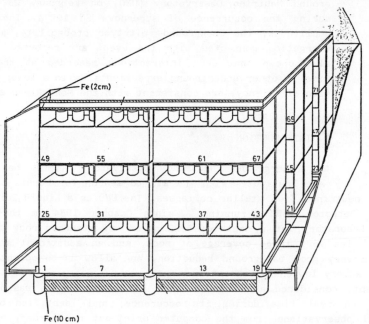

Fig.1 - The 90 ton Liquid Scintillation Detector in the Mt.Blanc Lab.

C_nH_{2n+2} (with $\bar{n}$ = 10), containing $8.4 \; 10^{30}$ free protons. The low energy local radioactivity background from the surrounding rock has been reduced by shielding each scintillation counter and the whole detector with more than 200 tons of Fe slabs.

In underground experiments, the main source of background is due to cosmic ray muons and their interactions in the rock surrounding the detector, which may induce contained pulses from secondary neutrons or gamma rays. This source of background is very low at the large depth underground of the Mont Blanc Laboratory, where about 3.5 muons per hour are recorded on the average in the whole LSD detector.

The liquid scintillator is watched from the top of each counter by 3 photomultipliers, in a 3-fold coincidence within 150 ns. Our calibrations[1], both from muons and with a ^{252}Cf source, show that a 1 MeV energy loss yields on the average 15 photoelectrons in 1 scintillation counter.

The low background of the LSD experiment and the high energy resolution of the scintillation counters allow us to operate the UNO at very low energy thresholds: fig. 2 shows the LSD detection efficiency

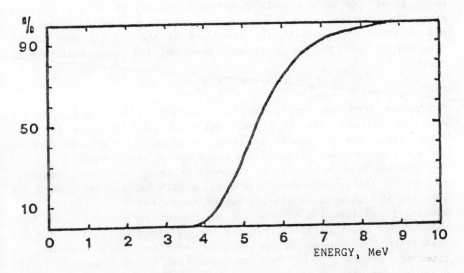

Fig. 2 - The LSD detection efficiency

as a function of the visible energy of the detected pulses. The electronic system consists of 2 levels of discriminators for each scintillation counter. A high-level (HEL) discriminator for pulses above the energy threshold $\sim$ 6-7 MeV for the 56 surface counters, and $\sim$

5 MeV for the 16 internal ones, with a total trigger rate of 0.012 Hz. A low-level (LEL) discriminator for pulses above the energy threshold 0.8 MeV is active only during a 500 μs wide gate, opened for all the 72 counters by the main high-level trigger. Two ADCs per counter measure the energy deposition in the scintillator in 2 overlapping energy ranges. A TDC gives the relative time of the interactions with a resolution of 100 ns. Three memory buffers, 16 words deep, for the 2 ADCs and the TDC of each scintillation counter, allow us to record all pulses without dead time. On-line software prints any burst of pulses satisfying our operational definition of a neutrino burst, namely a burst of pulses above a given multiplicity in a given time.

This recording system allows us to detect both products of $\bar{\nu}_e$ interactions with the free protons of the scintillator (namely, positrons and gammas in a delayed coincidence within 500 μs), through the capture reaction:

$$\bar{\nu}_e + p \rightarrow n + e^+$$
$$\hookrightarrow n + p \rightarrow d + \gamma$$

which gives the main signal in detecting neutrinos from collapsing stars. In addition, also electrons from elastic scattering of neutrinos of other species with the electrons of the scintillator can be detected in the LSD. For positron detection, the pulse amplitude is given by:

$$E_e = T_e + m_e c^2 \simeq E_\nu - 1.3 \text{ MeV}$$

The gammas from the $(np, d\gamma)$ capture reaction, with energy $E_\gamma = 2.2$ MeV, are emitted with an average delay of ~ 200 μs after the main interaction; the efficiency to detect a gamma, in the same counter where the neutron was produced, is about 40% on the average.

The absolute time in the LSD is recorded by using the signal broadcasted by the Italian Standard Time Service (IEN Galileo Ferraris). The accuracy of the absolute time is better than 2 msec.

The energy calibration of the LSD experiment is made by using both the 2.2 MeV peak of gammas from the $(np, d\gamma)$ capture reaction and the distribution of energy losses of near vertical muons crossing the detector, peaked at ~ 150 MeV. Neutrons from the spontaneous fission of a low activity ^{252}Cf source, placed in different positions inside a scintillation counter, are used as a neutron source for calibrating the detector. The accuracy of both calibration techniques is about ± 10%, and is limited at low energies by the ADCs resolution, and at high energies by the low muon statistics (on the average, 3.5 muons per hour cross the LSD experiment). The photomultipliers amplification and their stability are checked up about once per month, by using a ^{60}Co source placed onto the scintillation counter.

3. THE NEUTRINO BURST DETECTED IN THE LSD EXPERIMENT ON FEB. 23.12

Since January 1, 1986, the LSD experiment has been running with an average efficiency of 90% (and almost 99% since October 1986). Recently, the detector shielding has been partly increased for test purposes, with paraphin and lead in order to further decrease the low energy background from the surrounding rock. Trigger pulses

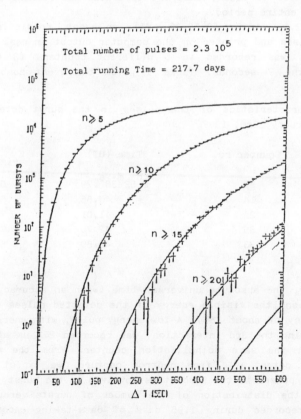

Fig.3 - Experimental distribution of bursts of pulses with multiplicity above 5, 10, 15 and 20 as a function of their duration (sec), with a binning of 10 sec, measured during 217.7 days.

are analysed in order to have a long term statistics and to search for bursts. The experimental time distributions of these pulses, grouped in bursts above a given multiplicity, are plotted in fig. 3 as a function of their duration and with a bin width of 10 sec. The distributions of fig. 3 refer to a data-taking period of 217.7 days (from September 28, 1986, to May 23, 1987). The smooth behaviour as well as the agreement with the predicted Poisson distributions (represented by the continuous curves) show that the trigger counting rate is stable and the detector is properly working during this time. For monitoring purposes the detector counting rate is also checked on-line every 100 triggers. By analysing these data, and from the tape analysis several days before and after the event discussed here, we have been able to verify that the apparatus was running properly throughout the entire period.

On February 23.12, 1987, ($2^h 52^m 36^s$ UT), an event, consisting of a burst of 5 pulses and printed on the computer output in real time at the occurence, was recorded in 5 different counters (3 of them internal) during 7 seconds. Table I gives the event number, the

Table I - Characteristics of the pulses in the burst detected on February 23rd, 1987

Event no.	Counter no.	Time (UT)	E_{vis} (MeV)
		$2^h 52^m 36^s.79$	
994	31		6.2
995	14	40.65	5.8
996	25	41.01	7.8
997	35	42.70	7.0
998	33	43.80	6.8

counter number, the absolute universal time (with an accuracy better than 2 msec), and the visible energy of the detected pulses (with an average accuracy of about 15%). A low energy pulse, with energy $E = 1$ MeV, accompanying the 3rd interaction, was recorded 278 μs after the main pulse in the same scintillation counter. From the measured efficiency to detect γ's from neutron capture, we expect on the average 2 such pulses in the 5 counters involved in the burst.

Fig. 4 shows the distribution of the number of bursts versus their multiplicity recorded during 1.96 days of data-taking encompassing the event discussed here. The full lines are computed according to a Poisson distribution of the trigger counting rate, with a binning of 10 seconds, and mean value given by our average raw trigger rate,

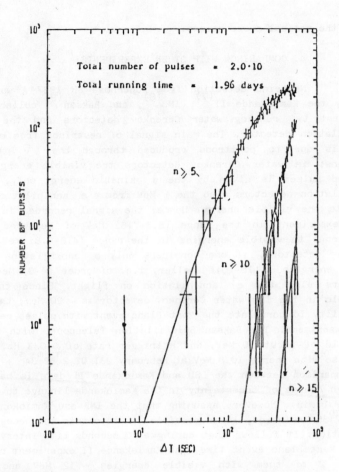

Fig.4 – The same as for fig.3, but for the run (of duration 1.96 days) including the event of Feb. 23, 1987, shown by a large dot.

which has the stable value of 0.012 Hz during this run as for the previous ones. From fig. 4 excellent agreement between the expected and measured distributions is found, except for the point at $t = 7$ sec, which has been added just to show the event considered here. The imitation rate of this event, evaluated in the on-line analysis from the raw trigger rate, is 0.7 per year, or $6.5 \; 10^{-4}$ in the time interval corresponding to the uncertainty of the instant of collapse, i.e. the 8.2 hours interval from February 23.08 to 23.44, as suggested by the optical observations.

4. COMPARISON WITH OTHER EXPERIMENTS

Neutrino detection from supernova SN 1987 A was reported also by the Kamiokande II[7], IMB[8], and Baksan[9] collaborations, the first two running water Cerenkov detectors and the third a scintillation detector. The main signal of neutrinos from collapsing stars is due to positrons produced through the ($\bar{\nu}_e$,p) capture reactions; in water Cerenkov detectors the visible energy of the recorded pulses is given by the e^+ kinetic energy only, while in scintillation detectors also the 1 MeV from e^+e^- annihilation contributes to the visible energy. Hence, the signal recorded in the Mont Blanc experiment in the range (5.8-7.8) MeV of measured energies corresponds to visible energies in the range (4.8-6.8) MeV in water Cerenkov detectors, if one considers only e^+ annihilation at rest, and at energies ~ (10-15)% smaller, i.e. of order (4-6) MeV, if one considers also the e^+ annihilation on flight. Since the energy threshold in the IMB water Cerenkov detector is ~ 20 MeV, there is no possibility to correlate the Mont Blanc event with pulses recorded in this experiment. The Baksan Scintillation Telescope, with an energy threshold of about 10 MeV, and a trigger rate of 0.033 Hz, detected one pulse with energy 10.8 MeV at February 23, UT $2^h 52^m 34^s \pm$ 2 sec.
The comparison between the LSD and Kamiokande II data is hampered by the high degree of uncertainty in the Kamiokande linkage to universal time ($\pm$ 1 min). However, assuming that the IMB and Kamiokande events at UT $7^h 35^m$ coincide, it is possible[10] to obtain the necessary time correlation. It follows that during a 10 seconds time interval closed to the Mont Blanc event time, the Kamiokande II experiment recorded 4 pulses, 2 of them with visible energies ~ 12 MeV and ~ 8 MeV respectively, energies which are considered significative in the signal detected at $7^h 35^m$ UT. It has been shown[6] that the Mont Blanc-Kamiokande events are not contradictory from the experimental point of view, and the two signals fit[10,11] the same model of a stellar collapse, emitting a high luminosity burst of low energy neutrinos.
The combined probability of the Mont Blanc-Kamiokande events gives a random coincidence rate of one every at least 77 years[12], or to a rate of less than 1.2 10^{-5} random coincidences in the 8.2 hours uncertainty in the time of the collapse.
Finally, close to the Mont Blanc event time, also the 2 running gravitational wave antennas, operating in Rome and Maryland at room temperatures, detected[13] a signal not due to seismic noise.

5. DISCUSSION

The energy spectrum of the neutrinos emitted from a collapsing stellar core can be approximated$^{(14)}$ by a distribution similar to a Fermi Dirac one, namely:

$$\Phi(\bar{\nu}_e/\text{sec MeV}) \propto \frac{\varepsilon^2 e^{-\alpha\varepsilon^2}}{1 + e^\varepsilon} \qquad (\varepsilon = E_\nu/KT)$$

where E_ν is the $\bar{\nu}_e$ energy (in MeV), and T the temperature of the neutrinosphere. The correction factor $\exp(-\alpha\varepsilon^2)$ takes into account neutrino absorption in the stellar envelope above the neutrinosphere. The central temperature KT of a type II presupernova is supposed to be less than 1 MeV, and the temperature of the neutrinosphere, after neutrino trapping, is of the order of a few MeV. Depending on the values of the absorption parameter α, the energy spectrum of the neutrinos emitted from the collapsing stellar core can be different from the spectrum produced during the collapse. Because of the energy dependence of the neutrino cross section, high energy neutrinos may be more absorbed than the low energy ones, and bigger the mass of the star envelope stronger is the shift of the spectrum towards low energies. The neutrino interactions recorded in the LSD experiment agree with a temperature of the neutrinosphere of ~ 2 MeV and absorption parameter $\alpha \sim 0.1$.

The total energy involved in the burst can be estimated assuming that the 5 pulses recorded in the LSD are due to $\bar{\nu}_e$ capture processes. This assumption seems natural as the $\bar{\nu}_e$ capture cross section is about 2 orders of magnitude higher than the scattering cross section of ν_e with the electrons of the scintillator. Adopting a distance of 52 kpc for SN 1987A, and using the cross section:

$$\sigma(E_\nu) = 9.45 \cdot 10^{-44} (E_\nu - \Delta M) \left[(E_\nu - \Delta M)^2 - m_e c^2\right]^{1/2} \text{cm}^2$$

where ΔM is the neutron to proton mass difference, and m_e is the electron rest mass, the total energy recorded in the LSD experiment is $\sim 6 \cdot 10^{53}$ erg. The total energy outflow in $\bar{\nu}_e$, integrated over the energy spectrum (3), is:

$$1.2 \cdot 10^{54} \lesssim E_\nu \lesssim 6.3 \cdot 10^{54} \quad \text{erg} \qquad (\text{at 90\% c.l.})$$

The limited statistics of the number of pulses in the burst may explain why this value is slightly higher than the predicted theoretical expectation.

The expected number of $\bar{\nu}_e$ interactions in the other underground experiments can be estimated by using their experimental detection efficiency and the energy spectrum (3). Table II gives the expected and recorded number of pulses in the experiments which reported neutrino detection from SN 1987 A. In spite of the larger number of target protons in comparison with the LSD, the number of recorded interactions is rather small because of the lower detection efficiency of these experiments at low energies. Hence, from the experimental point of view, there is no contradiction between the event detected in the Mont Blanc Neutrino Observatory and in the other detectors at the same time.

Table II - Expected and detected $\bar{\nu}_e$-interactions in other experiments.

Experiment	Expected	Detected
BAKSAN	$1.1 \pm .5$	1
KAMIOKANDE	$\leq 10 \pm 8$	2
IMB	0	0

6. CONCLUSIONS

The neutrino burst recorded on real time in the Mont Blanc Underground Neutrino Observatory, during the occurence of supernova SN 1987 A, is self consistent and indicates that the collapse had a duration of several seconds, during which a high luminosity burst of neutrinos was emitted by a low temperature neutrinosphere. Even if at large distance, this stellar collapse produced a significant signal in the LSD experiment because of its very low energy threshold and high efficiency to detect low energy pulses. A second burst, detected 4.7 hours after the Mont Blanc one, is not contradictory from the experimental point of view, and indicate that the stellar collapse in the Large Magellanic Cloud developed on at least two stages.
This conclusion is further supported by the luminosity and spectral evolution of the supernova at the very early stages, which agrees[15] with the start time of the collapse as given by the Mont Blanc event time. Neutrinos detected at $7^h 35^m$ UT may have been emitted in a delayed pulse from the neutron star already existing. Also the presence of a very bright companion star[16,17], caused by this supernova explosion, is hard to be explained within the framework of the standard theoretical models, based on spherical simmetric collapse, without angular momentum and magnetic field. The two bangs scenario[10,18,19] seems to fit better the experimental evidences availa-

ble; this scenario can be a simple, natural explanation of the two neutrino pulses, which, because of the different energy spectra, have induced different information in detectors of different type.

REFERENCES

1. G.Badino et al., Nuovo Cim.,7C,573,1984
2. M.Aglietta et al., Nuovo Cim.,9C,185,1986
3. M.Aglietta et al., IAU Circ.no.4323 (Feb.28, 1987)
4. M.Aglietta et al., Europhys.Lett.3,1315,1987
5. V.L.Dadykin et al., JETP Lett., (in russian), 45,464,1987
6. M.Aglietta et al., Europhys.Lett., 3,1321,1987
7. K. Hirata et al., Phys.Rev.Lett., 58,1490,1987
8. R.M.Bionta et al., Phys.Rev.Lett., 58, 1494, 1987
9. L.N.Alexeyeva et al., JETP Lett. (in russian), 45, 461, 1987
10. A. De Rujula, Phys. Lett. B, 193, 514, 1987
11. D.N.Schramm, Comments on Nuclear and Particle Physics, 1987, in press
12. A. De Rujula, CERN TH-4839, 1987
13. G.Pizzella et al., Proc. 4th G.Mason Workshop in Astrophysics, Fairfax, 1987, in press
14. D.K.Nadjozhin, I.V.Ostroschenko, Sov. Astron., 24, 47, 1980
15. E.J.Wampler et al., Astr. and Astrophys. Lett., 1987, 182, L51
16. C.Papaliolios et al., Proc. 4th G.Mason Workshop in Astrophysics, Fairfax, 1987,in press
17. W.P.S.Meikle et al., Nature, 1987, in press
18. W.Hillebrandt et al., Astron. Astrophys. Lett., 1987, 180, L20
19. L.Stella and A.Treves, Astron. Astrophys. Lett., 1987, in press

SEARCH FOR NEUTRINO SOURCES WITH THE FREJUS DETECTOR *

HINRICH MEYER
UNIVERSITY OF WUPPERTAL

INTRODUCTION

Proton decay searches with sophisticated detectors deep underground open up the possibility to look for point sources of neutrinos. The existence of specific astronomical objects as sources of high energy neutrinos (of the order of TeV) has been discussed since more than 20 years [1]. Supernova explosions are among the most prominent possiblities and SN1987A in the LMC provides a unique case to test the various predictions [2]. Furthermore the open question of the nature of dark matter asks for explanations on the basis of new not yet discovered elementary particles. In recent years several particles predicted to exist in supersymmetric theories have been proposed as candidates for the dark matter [3]. If gravitationally collected in sufficient numbers in the sun (earth) they may annihilate to normal matter and produce detectable neutrino fluxes, [4] surely a very spectacular possibility for neutrino astronomy. Significant numbers of neutrino events have been recorded in the proton decay detectors, by observing muon- and electron-neutrino interactions inside their fiducial volume and through the detection of muons from charged current muon-neutrino interactions entering the detector from below the horizon. The neutrinos originate from the interactions of cosmic ray nuclei with the nuclei of the earth atmosphere and have a nearly isotropic distribution [5]. These neutrino events provide a natural background level and the data basis of the search for neutrinos sources.
The data described in this paper have been taken with the FREJUS-detector continuously since Feb.1984. At that time the detector had a size of ~240 tons and grew linearly to its final size of ~900 tons by June 1985. The main mass of the detector consists of 3 mm iron plates 6m x 6m in area interleaved with flashtubes (~940.000) in two orthogonal views and (~40.000) Geigertubes used to trigger the detector. The trigger rate is (40-50)/h, about half of them due to random coincidences. The rate of cosmic ray muons is 20/h, however for neutrino events with the vertex inside the detector only about 2/week. More details about the experiment can be found in Ref.6.

1. HIGH ENERGY NEUTRINOS FROM SN1987 A

As early as 1934 W. Baade and F. Zwicky [7] proposed that supernovae explosions are the main source of cosmic rays. A general discussion of acceleration processes and particle fluxes from young supernovae can be found in Ref.8 and 9. They specifically consider the case of "small" neutrino detectors that have been operational for some time mainly to search for proton decay. Soon after the discovery of SN1987A, Gaisser and Stanev [10] and Sato [11] gave estimates for neutrino fluxes from the supernova pointing out that presently running detectors may indeed have enough sensitivity to observe a high energy neutrino flux from the supernova. The neutrinos are supposed to originate from very high energy proton interactions in the supernova shell producing charged pions that decay to muons and neutrinos. Similarly a high energy photon flux is to be expected from the decay of neutral pions. It has been emphasized that proton energies up to 10^{16} eV [12] may be reached with a power law energy spectrum $\sim 1/E^n$ and with a spectral index n as low as 2. The time scale for the acceleration process to become effective is estimated to be of the order of month after the SN event however it may well take a much longer time [13].

High energy neutrinos (ν_μ, $\overline{\nu_\mu}$) are observed through charged current production of muons which at high energies have ranges in the rock around the detector of order a few km [14]. The muon energy loss is to a good approximation given by

$$dE/dx = a + b \times E$$

with a $\approx$ 0.22 GeV/ mwe and b $\approx$ 4.4 x 10^{-4} / mwe and m w e (= meter water equivalent) is the matter thickness in units of 100 g/cm². The mean muon energy to penetrate a depth L [mwe] in terms of the parameters a,b then is

$$E = \frac{a}{b} \left(e^{b \times L} -1 \right)$$

with a/b = 480 GeV and 1/b = 2.2 x 10^3 mwe.
The probability to observe a muon from a ν_μ -interaction in the surrounding rock for neutrino energies up to 10^{20} eV has been calculated by several authors [14] taking account of the improved knowledge of nucleon structure functions. A ν-source with an integral energy spectrum

$$I (> E) \sim 1/E$$

gives at the detector a very high average muon energy of order a few TeV [14]. Muons of such a large energy have large radiative energy losses in the detector, e.g. for a muon energy of 5 TeV in the Frejus-detector structure (2g/cm² Fe) one has

$$2 \, mwe/m \times 4.4 \times 10^{-4}/_{mwe} \times 5000 \text{ GeV} \approx 4.5 \text{ GeV/m}$$

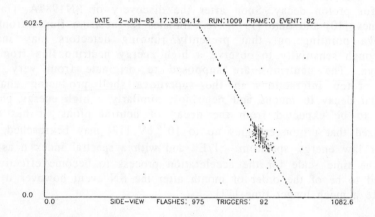

Fig 1: A high energy muon in the FREJUS-detector. An electromagnetic shower of ~30 Gev, produced by a bremsstrahlungsphoton, an electron positron pair or a knock-on electron has developed about 5 meters from the muon entrance point.

This energy loss is easily recognized in the FREJUS detector due to the very fine granularity, an example is shown in **Fig. 1**. The probability to observe muons from high energy supernova neutrinos depends strongly on the spectral index n of the proton energy distribution at the supernova as can be seen from **Fig. 2** which shows the integral number of muons (with energy higher than E) in a detector of 100m² area for one year as function of the spectral index n. The source is assumed to be at a distance of 10 kpc and with a proton luminosity of 10^{43} erg/sec. In the Frejus detector upwards muons from ν´s can in general not be distinguished from downgoing atmospheric muons only if they either come to rest in the detector (stopping muon, using multiple scattering information) or if they have sufficiently large radiative energy loss.(see Fig.1) In the latter case the pattern of the shower development gives the directional information.

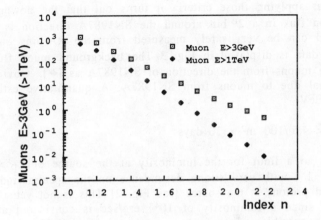

Fig 2: The number of muons with energy greater than E from neutrino interactions pointing to the detector for two threshold muon energies (3 GeV and 1000 GeV). A proton source with energy spectrum $1/E^n$ is assumed with power output 10^{43} erg/sec at a distance of 10 kpc. The detector area is 100 m^2.

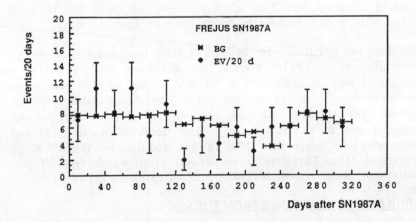

Fig 3: The number of muons in an angular cone of 2° opening at the position of SN1987A as observed in the FREJUS-detector starting at the time of the supernova on 23 Feb. 1987 averaged over 20 days. The muons all enter the detector from above and therefore are background to a possible signal from SN1987A. The expected background level is calculated from the average muon flux in 20 RA bins outside the position of SN1987A. It is not constant with time because of various off-times of the detector.

But even without applying those criteria it turns out that the downgoing atmospheric muon flux in a 2° bin around the SN1987A direction is only ~ 7/20 days and can be very safely measured from the RA-bins outside SN1987A. Our data is displayed in Fig. 3. The background muon flux is shown as [x] the muons from the direction of SN1987A as [♦]. Obviously there is no signal due to muons from SN1987A. A quantitative estimate gives for $E_\mu > 3$ GeV

$$N_\mu < 4.5 / 100 \text{ m}^2 \times 20 \text{ days}$$

This corresponds to a limit for the luminosity at the source of $L_p > 2 \times 10^{43}$ erg/sec [n=2] with very strong dependence on the spectral index n of the accelerated protons at the source [see Fig. 2] This is not yet a very significant limit since a luminosity of 10^{43} erg/sec is considered at the upper end of the possible values for a young supernova [9,13]. It seems however not unreasonable to assume that the luminosity is not smooth in time rather we would expect to see a strongly fluctuating ν-flux at least just after the neutrino flux becomes detectable. Then on short time scales the neutrino flux could be rather large without violating energy bounds at the source. Since this kind of measurement can be performed with other detectors as well (Baksan, Kamiokande, Kolar, Nusex, IMB), a combined detection area from all experiments of about 800 m² could be reached with a correspondingly better limit on neutrino fluxes.

Detection of neutrinos do have a rather low sensitivity to proton acceleration in SN1987A because the detection probability for muon neutrinos is only of order 10^{-5} at 10 TeV. A flux of high energy photons from neutral pion decay on the other hand, (detectable only in the southern hemisphere) will be reduced by γ γ absorption in the supernova due to a rather large column depth of matter [15] and for photon energies larger than 100 TeV in addition by the 3° K photon background [16]. Therefore it is not yet clear which type of particle, neutrinos or photons from SN1987A will be easier to detect.

2. HIGH ENERGY NEUTRINOS FROM THE SUN

Dark matter seems to dominate over luminous matter seen as stars on all scales in the universe [17]. Elementary particles with new weak interactions and produced at early times of the big bang have been proposed as candidates for the dark matter [18]. For the special case of supersymmetric particles very interesting signals based on observing photons or antiprotons from the annihilation of dark matter particles in the galaxy have been proposed [19]. Of particular importance however

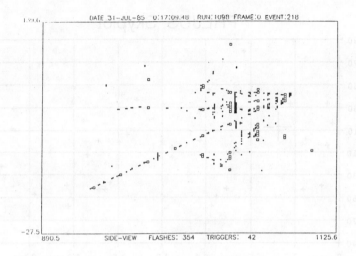

Fig. 4: A high energy muon-neutrino event from the FREJUS detector

for the underground proton decay experiments are predictions for measurable fluxes of high energy neutrinos from the sun (earth) [20]. Dark matter particles having semi-weak interaction cross sections with normal matter may scatter inside the sun (or the earth) thus leaving their Keplerian orbits and enter closed orbits inside the sun [21]. This kind of process will then lead to an increasing abundance of dark matter particles in the sun and may have reached after 4,6 x 10^9 years the lifetime of the sun a fraction (by mass) of $\sim 10^{-12}$. It has been shown that for dark matter particle masses larger than $\sim$ 3,0 GeV the particles are retained in the sun [22]. Dark matter collisions inside the sun can lead to annihilations into normal matter (quarks and neutrinos). The direct neutrinos as well as neutrinos from (mostly) heavy quark decays are predicted to come with fluxes of high energy (> 1GeV) neutrinos close to the detection limit of the present generation proton decay detectors [23]. The neutrino events as observed in the FREJUS detector allow rather precise reconstruction of the direction of the incoming neutrino, we estimate a value of < 20° at visible energies > 2 GeV. The data sample reported here consists of 37 events with visible energy > 2 GeV. One event from this sample is shown in **Fig.4.** From the absolute time of occurence of the neutrino events in the detector and the neutrino direction the point of origin in the sky can be calculated. The resulting skyplot is shown in **Fig.5** using right ascention (RA) [h] and declination (δ) [°] as coordinates. The distribution is rather smooth and consistent with atmospheric origin of all observed neutrino events.

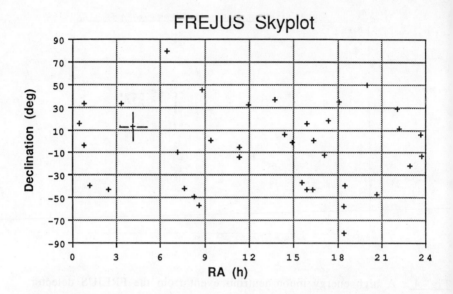

<u>Fig.5:</u> A skyplot of the high energy (> 2 GeV visible energy) neutrino events as observed in the FREJUS detector. The observation time is from Feb. 1984 - Jan. 1988. The average angular accuracy is indicated for one event in the plot.

For a determination of the background flux of high energy neutrinos from the sun the differences in right ascention [ΔRA] and declination [$\Delta\delta$] between the apparent point of origin of the neutrinos and the position of the sun is calculated and shown in **Fig.6**. The error box at the position of the sun corresponds to an estimated angular resolution of < 20°. No event is observed from the direction of the sun. The efficiency to observe a neutrino with energy E based on the visible energy E_{vis} in the FREJUS detector depends on the type of event, it is highest for charged current electron neutrino events and lowest for neutral current events. At the present stage of the analysis we estimate (for a visible energy cut E_{vis} > 2 GeV) the detection efficiency for neutrinos of energy 2 GeV to be about 50% and better values at higher energies. For an integrated luminosity of 1,6 kty we therefore estimate

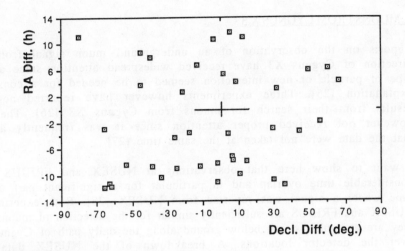

Fig6: The differences in RA and δ with respect to the position of the sun for the high energy neutrino events in the FREJUS detector. The error bars at the position of the sun (center of the plot) indicates the pointing accuracy achieved.

.n upper limit on the neutrino flux with E > 2 GeV of

$$N > \frac{2,3}{0,5} \times \frac{1}{1,6 \text{ kty}} = 3 \text{ events/kty}$$

This number puts interesting bounds on several supersymmetric particle candidates for the galactic dark matter. The most recent analysis of this situation can be found in Ref.[24]. It can be said that the supersymmetric dark matter scenario is not excluded. Certainly higher statistics data would be very useful to set better limits. Order of magnitude improvements can however only be obtained from next generation detectors.

3. MUONS FROM CYGNUS X 3

Reports on the observation of an underground muon signal from the direction of Cygnus X3 have received widespread attention since a new type of particle or new interaction seemed to be needed for a consistent explanation [25]. Three experiments however have reported negative results from their search for muons from Cygnus X3 [26]. That has however not received proper attention since it was frequently argued that the data were not taken at the same time [27].

I want to show here that observations at NUSEX and FREJUS have considerable time overlap and in particular for a significant part of the time where a signal was reported from NUSEX [28]. Both experiments NUSEX and FREJUS are sufficiently similar for the detection of muons also they are at similar depth below ground along the daily path of Cygnus X3 over the detector locations. A breakdown of the NUSEX data was available to us for different time periods [29] and is shown for 1984 in Fig.7. About half of the total signal of NUSEX is contained in the 1984 data in the phase bin from 0.7-0.8.

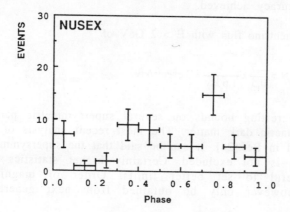

Fig. 7: The muon flux from the direction of Cyg-X3 observed with the NUSEX detector in 1984. The bin from 0.7 - 0.8 is the "signal" bin w.r.t. the 4.8 h phase of the source.

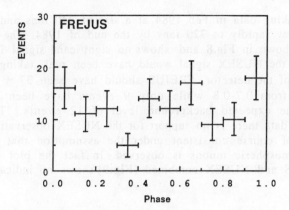

Fig.8: The muon flux from the direction of Cyg-X3 observed with the FREJUS detector in 1984. The higher rate observed by FREJUS is due to the larger size of the detector (~factor 3). No signal in any phase bin is observed.

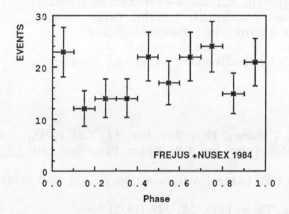

Fig.9: The combined data from NUSEX and FREJUS for 1984 from the direction of Cygnus-X3

FREJUS started taking data in Feb 1984 at a size of 240 tons and the siz of the detector grew rapidly to 720 tons by the end of 1984. The FREJUS data for 1984 is shown in **Fig.8** and shows no significant signal in any o the phase bins. If the NUSEX signal would have been real, taking accoun of the larger size of the detector, FREJUS should have seen 37 ± 8 event in the signal bin from 0.7-0.8 while only 9 events have been observe consistent with the expected background level of 12 events. Therefor from the FREJUS data there is no support for the NUSEX observation. Bot experiments are of course consistent under the assumption that only th background of atmospheric muons is observed. In fact the plot with th data from FREJUS and NUSEX combined [**Fig.9**] shows no indication of signal.

CONCLUSIONS

The neutrino events observed in the FREJUS detector are entirel consistent with atmospheric origin. No signal from either the direction c the sun or of Supernova SN1987A has been detected. The muon flu from the direction of Cyg X3 is completely understood as of atmospheri origin and therefore the observation of the NUSEX-collaboration is nc confirmed.

    *   Members of the FREJUS-Collaboration are:
       I. Physikalisches Institut der RWTH Aachen, Germany.
       Laboratoire de l´Accelerateur Lineaire, Orsay, France.
       LPNHE-Ecole Polytechnique Palaiseau, France.
       DPhPE-Saclay, Gif-sur-Yvette, France.
       Universität-Gesamthochschule Wuppertal, Germany.

REFERENCES:

1. T.K. Gaisser, T. Stanev, Phys. Rev. Lett. **54**, 2265 (1985)
   E.W. Kolb, M.S. Turner and T.P. Walker, Phys. Rev. Lett. **D32**,1145 (1985)
   C. Cesarsky, P.O. Lagage, Astron. Astrophys. **125**, 249 (1983)

2. H. Sato, Prog. Theor. Phys. **58**, 549 (1977)

3. D.N. Schramm, Nucl. Phys. **B252**, 53 (1985)

4. J. Silk, K.A. Olive and M. Srednicki, Phys. Rev. Lett. **55**, 257 (1985)

5. T.K. Gaisser, Todor Stanev, S.A. Bludman and H. Lee, Phys. Rev. Lett. **51**, 223 (1983)

6. Frejus-Collaboration, Nucl. Instr. and Meth. A262, 463 (1987)

7. W. Baade and F. Zwicky, Phys. Rev. (L) 46, 76 (1934)
    W. Baade and F. Zwicky, Phys. Rev. 45, (A), 138 (1934)

8. V.S. Berezinsky and O.F. Prilutsky, Astron. Astrophys. 66, 325 (1978)

9. V.S. Berezinsky, C. Castagnoli and P. Galeotti, IL Nuovo Cimento 8C, 185 (1985)

10. T.K. Gaisser and T. Stanev, Phys. Rev. Lett. 58, 1695 (1987) and 59, 844 (E) (1987)

11. H. Sato, Mod. Phys. Lett. A2, 801 (1987)

12. V. S. Berezinsky and V.L. Ginsburg, Nature 329, 807 (1987)

13. T.K. Gaisser, Alice Harding and T. Stanev, Nature 329, 314 (1987)

14. M.H. Reno and C. Quigg, Phys. Rev. D37, 657 (1988)
    T.K. Gaisser and A.F. Grillo, Phys. Rev. D36, 2752 (1987)

15. R.J. Protheroe, Nature 329, 135 (1987)

16. I.A. Bond et al., Phys. Rev. Lett. 60, 1110 (1988)
    I.A. Bond et al., KEK-Preprint 88-7

17. S.M. Faber and J.S. Gallagher, An. Rev. Astron. Astrophys. 17, 135 (1978)

18. J. Ellis et al., Nucl. Phys. B238, 453 (1984)

19. F.W. Stecker, S. Rudaz and T.F. Walsh, Phys. Rev.Lett. 55, 2622 (1985)

20. J. Silk and M. Sredznicki, Phys. Rev. Lett. 53, 624 (1984)
    A. Gould, Ap. J. 321, 560 and 571 (1987)

21. J. Faulkner and R.C. Gilliland, Ap.J. 299, 994 (1985)
    W.H. Press and D.N. Spergel, Ap.J. 296, 679 (1985)
    D.N. Spergel and W.H. Press, Ap.J. 294, 663 (1985)

22. K. Griest, D. Seckel, Nucl. Phys. B283, 681 (1987)

23. T.K. Gaisser, G. Steigman and S.Z.Tilav, Phys. Rev. D34, 2206 (1986)
    J.S. Hagelin, K.-W. Ng and K. Olive, Phys. Lett. B180, 375 (1987)

24. J. Ellis, R.A. Flores and S. Ritz, Phys. Lett. 198B, 393 (1987)

25. K. Ruddick, Phys. Rev. Lett. 57, 531 (1986)

26. E. Aprile et al., Proc. of the Int. Europhysics Conf. on High-Energy Physics p.424, Bari July 1985
    Y. Oyama et al., Phys. Rev. Lett. 56, 991 (1986)
    Ch. Berger et al., Phys. Lett. B174, 118 (1986)

27. F. Halzen, K. Hikasa and T. Stanev, Phys. Rev. D34, 2061 (1986)

28. G. Battistoni et al., Phys. Lett. B155, 465 (1985)

29. B. D'Etorre-Piazoli, private communication

ON THE POSSIBILITY OF DETECTING SOLAR AND SUPERNOVAE NEUTRINOS WITH In^{115} DETECTOR

A. K. Drukier

Applied Research Corporation
8201 Corporate Drive, Suite 920
Landover, MD 20785

and

Harvard-Smithonian Center for Astrophyics
60 Garden Street
Cambridge, MA 02138

ABSTRACT

The inverse β^- decay of In^{115} provides a low-threshold direct-counting neutrino detector. Unfortunately, the radioactive background due to In^{115} β-decay may be an obstacle in the development of a detector for pp solar neutrinos. We suggest that this background can be efficiently suppressed in searches for higher energy solar neutrinos, e.g. from Be^7, p-e-p and B^8. A signal/background ≈ 10-30 is expected. It may be possible to have a count rate of 0.2 events/day with a detector consisting of 10 tons Indium and about 15 m^3 of liquid Xenon. Unfortunately, pp neutrinos cannot be detected in this set-up.

Furthermore, the In^{115} detector will provide unique information about supernovae explosions, especially in our Galaxy.

INTRODUCTION

The detection and measurement of the neutrinos from the thermonuclear reactions providing the energy of the sun is a longstanding experimental problem[1].

The two best-known solar neutrino detectors are radiochemical[1,2,3] and based on Cl^{37} and Ga^{71}, respectively. Recently, a few new possibilities of detecting solar neutrino have been suggested. Three of them are based on neutrino capture in In^{115} (5,6,7). Furthermore, it may be possible to use coherent scattering on heavy materials[8]. The expected capture rates for the standard solar neutrino model are given in Tab. 1; all capture rates are in SNU units. In Tab. 2 we give the estimated mass of the detector which leads to a count rate of 0.2 counts/day. It should be pointed out that only the Cl^{37} detector is in operation and the Ga^{71} detector is practically ready for use, whereas the other detectors are in the development stage.

The experiment[2] (inverse beta decay in Cl^{37} leading to Ar^{37}) has yielded a result that is generally accepted to be positive but smaller than the prediction of standard solar models by a factor three. This detection scheme is, however, sensitive to neutrinos from B^8 which constitute only about 5×10^{-5} of the total flux and arise from a nucleosynthesis which occurs only in the hottest part of the solar interior. The calculation of the B^8 flux thus depends on a thorough understanding of the energy tranport in the sun as well as the knowledge of all the cross-sections involved in the chain of synthesis[1]. Furthermore, the discrepancy can be attributed to neutrino oscillation[4].

An experiment to measure the actual energy spectrum of neutrinos by the electronic detection of conversion events will provide information of considerable interest for both astrophysics and elementary particle physics:
- in astrophysics it is a very effective way to study the interior of the sun;
- in elementary particle physics an experiment with solar neutrinos will be extremely sensitive to neutrino oscillations due to the enormous distance between the sun and the earth and low energy of these neutrinos. It will be possible to search for mass differences down to the order of 10^{-6} eV.

The idea of using Indium for detecting solar neutrinos from the proton-proton primary reaction in the sun has been suggested by R. S. Raghavan. The basic reaction is the β-inverse decay of In^{115} to an excited state of Sn. It was soon recognized[5] that the major problem of the experiment is the radioactive β-decay of In^{115}. To date, three different proposals of neutrino detector In^{115} detector were published[5,6,7]. The physics and state of the art of In^{115} neutrino detector is discussed in section 1.

It may be easier to build larger but much simpler detectors of Be^7, p-e-p and B^8 solar neutrinos when using In^{115}. In chapter 2 we discuss the problem of background suppression for higher energies, E > 0.5 MeV neutrinos and the tentative implementation which requires 10 tons of In, 15 m^3 of liquid Xenon and 10^5 electronic channels.

1. **In-115 as solar neutrino detector**

The basic reaction is the β inverse decay of In^{115} to the 614 keV-excited state of Sn:

$$\nu_e + \text{In} \rightarrow \text{Sn}^{**}(7/2^+) + e^- \tag{1}$$

This reaction (1) has a threshold of 120 keV so that:

$$E_e = E_\nu - 120 \text{ keV}$$

Furthermore, it is important that the prompt electrons are emitted in the forward direction, i.e., that there is a correlation between the direction of the neutrino and the electron.

Reaction (1) is followed by a double de-excitation of Sn^{**}

$$\text{Sn}^{**}(7/2^+) \rightarrow \gamma(100 \text{ keV}) + \text{Sn}^*(3/2^+) \tag{2}$$

$$\text{Sn}^*(3/2^+) \rightarrow \gamma(500 \text{ keV}) + \text{Sn}(1/2^+) \tag{3}$$

The half life of $\text{Sn}^{**}(7/2^+)$ is 3.26 μs. The half life of $\text{Sn}^*(3/2^+)$ is less than 10^{-10} sec. The 100 keV and 500 keV photons are emitted isotropically. In reaction (2) the photon's internal coversion probability is 50%.

The cross section for the neutrino capture is given by

$$\sigma = \frac{2\pi^2}{c} \times \frac{n}{m_e c} \times \frac{\ln 2}{(ft)_\beta} \times W_e^2 \times G(Z, W_e, p_e) \tag{4}$$

with W_e being the total energy (including rest-mass) of the electron following neutrino capture: $W_e = E - Q + 1$, E being the energy of the solar neutrino and Q the threshold energy, all in units of $m_e c^2$. $G(Z, W_e, p_e)$ is the modified Fermi function[9] depending on the charge number of the final nucleus <Z(Sn) = 50> as well as the energy and momentum p_e (in units of m_ec) of the emitted electron. The value for $(ft)_\beta$ was estimated to be 2.5×10^4 sec (see ref. 5).

The electron energy spectrum obtained from folding the solar neutrino spectrum with the cross sections of reaction (1) is given in Tab. 1. A ν_e + In reaction is mainly sensitive to pp and Be^7 neutrinos and will then give information complementary to the Davis experiment. For pp neutrinos, In^{115} has a very large capture rate, ≈ 700 solar neutrino units (SNU's) [1 SNU = 10^{-36} captures/(target nucleus) · sec]. It also has a delayed-coincidence signature for neutrino events due to the microseconds life of Sn^{115**}. Taking into account the fact that In is almost monoisotopic the amount of material needed for an experiment with an event rate of 1/day is only about 3 tons. With exception of detectors based on coherent scattering[8], this is the smallest amount of material possessing natural isotopic abundance which has so far been proposed to yield this given solar-neutrino capture rate.

Unfortunately, there is a radioactive background due to

$$^{115}In \rightarrow {}^{115}Sn + e^- + \bar{\nu}_e \qquad (5)$$

The half life of In^{115} is 5×10^{14} years. The maximum energy of the electron is 490 keV. It is clear that unless great care is taken, the accidental coincidence rates could make the experiment unfeasible. These features have been discussed by R. S. Raghavan[5] and by Booth[10]. The background is a most serious problem to overcome since low-energy radiations must be detected. A possible solution is indicated by the fact that in addition to constraints on energies, pulse shapes, and delay time, it could also be demanded that the two pulses originate in the same location of the neutrino detector. The comparison of diverse inverse β^- neutrino detectors is provided in Table 2.

In 1984 we suggested[11] another possibility: liquid Xenon Multiwire Proportional Chamber (MWPC) with indium convertor. For p-e-p, Be[7] and B[8] neutrinos and using thin In foil claded with plastic it is possible:
- absorb all β-particles inside the plastic cladded In-converter;
- permit escape of the majority of prompt electrons from the In-converter;
- have most of 500 keV photons absorbed in liquid Xenon.

2. Use of the particle range as means of background rejection

In the following we will show that one can build a neutrino detector which consists of only MWPC's and an indium converter. Unfortunately, using this method one cannot detect pp neutrinos.

We require only the double coincidence of the prompt electron ($E_e=E_\nu-120keV$) and of photoelectron from the absorption of 500 keV photon. The simplified detector configuration is shown in Fig. 1. The liquid Xe MWPC chambers are used as both photon converter and a high spatial resolution detector of prompt electrons. They have:

a) good spatial resolution of say, a few mm^3;
b) good timing resolution of a few microseconds;
c) a high absorption efficiency and reasonable energy resolution, $\Delta E/E(500~keV) \approx 25\%$.

A good spatial resolution of liquid Xenon MWPC permits us to define a subclass of events close to the In convertors (zone 1). These are the electrons due to neutrino absorption and/or β-background. The 500 keV photons will be detected in a few centimeters of the converter in zone 2. The signature of the neutrino event is

a) one particle (electron?) in zone 1
   one particle (photon?) in zone 2;
b) particle in zone 2 no more than 10 μsec after the event in zone 1;
c) the energy deposed in zone 2 in the range of 400 < E < 600 keV.

The background can be due to the following combination of natural radioactivity inside of the detector: ββ, βγ, γγ. The background is studied in ref. 11 under the assumption that the natural β-activity of In^{115} dominates the electron background. The predicted backgrounds give for Be-7 neutrinos:

$$X_{\beta\beta} = \text{signal} / \beta\beta \text{ background} = 9 \times 10^{-5}/m = 2.7 \times 10^{-3} \quad (7)$$

$$X_{\beta\gamma} = \text{signal} / \beta\gamma \text{ background} = 1.1 \times 10^{-3} \quad (8)$$

$$X_{\gamma\gamma} = \text{signal} / \gamma\gamma \text{ background} = 54 \times m = 2.1 \quad (9)$$

where m = (mass of In/mass of liquid Xenon), say m = 0.04. To calculate the photon background, we assumed

$$n_\gamma = 10^{-8} \text{sec}^{-1} \text{ g}^{-1} \text{ keV}^{-1} \quad (10)$$

observed due to γ -emitting impurities (data of F. Rines for a plastic scintillator). Actually, it is possible that γ radioactivity of liquid Xenon is much lower than in a plastic scintillator. In the following we will assume that $X_{\beta\gamma} = 10^{-3}$ or 10^{-2} for $n_\gamma = 10^{-8}$ and 10^{-9}, respectively. A good solar neutrino experiment should feature a signal/background ≈ 10, i.e. order of magnitude suppression of radioactive background is necessary. For example, four orders of magnitude suppression of β-background and a factor three suppression of γ-background leads to $X_{\beta\beta} > 10^4$, $X_{\beta\gamma} \approx 30$ and $X_{\gamma\gamma} \approx 20$ (see analysis in ref. 11).

How to get suppression of β-background by 4 orders of magnitude? The main idea of the proposed set-up is very simple; we suggest the use of indium converter cladded with appropriate thickness of plastic. The thickness of the plastic cladding is bigger than the maximum range of β-particles (E_{max} = 490 keV R_{max} = 0.16 g/cm^2). At the same time, a fraction of prompt electrons escape. For electrons with E < 3 MeV, the range is given by[9]

$$R [g/cm^2] = 0.412 \, E^n \tag{11}$$

with

$$n = 1.265 - 0.0954 \ln E \tag{12}$$

where E is in MeV. In table 3 we show the range of electrons of interest. In figure 2, we show the range vs. energy for E_e < 1.5 MeV.

It should be pointed out that the background radiation would be almost perfectly eliminated. First, electron losses are given in this case by hundreds of independent interactions and for <u>thick target</u> Landau fluctuations are very small. Second, only a very small fraction of β's have an energy greater than

400 keV and less than 0.6% of β-particles have a range greater than 0.1 g/cm². Furthermore, the β's are emitted isotropically, i.e. most of them will cross much more than 0.16 g/cm². The formula to calculate the probability of electron escape from a cladded In converter are given in ref. 11. The data for an Indium sheet with a thickness of L = 0.5 x R_{max}, R_{max} and 1.5 x R_{max}, respectively, were calculated for diverse converters claded with 1.5 x R_{max} < d < 2.0 x R_{max} of plastic. For example, for Indium thickness of R_{max} = 0.16 g/cm², the probabilities of escape of β electrons are 10^{-4} and 10^{-5} for plastic thickness of 0.19 g/cm² and 0.22 g/cm², respectively. Thus, the β-background will be adequately well suppressed with the plastic cladding of ca. 1.5 R_{max} ≈ 0.24 g/cm²/.

It can be observed that use of a "claded Indium converter" improves only the $X_{\beta\beta}$ and $X_{\beta\gamma}$. Fortunately, in ref. 11 we assumed ΔE/E(FWHM) ≈ 25% whereas the recent progress in noble liquids TPC permits ΔE/E(FWHM) ≈ 5%. Thus, we expect

$X_{\gamma\gamma}$ = signal/γγ background ≈ 50   (13)

The detector efficiency can be estimated to be

QDE = QDE(γ) x (1-P_{In} (t)) x D_e (Ev, t, d)   (14)

where:

a) QDE(γ) ≈ 90% is only weakly dependent on the converter geometry and cladding (see ref. 11);

b) If the 500 keV photon is absorbed inside the In converter, then the photo-electron has an energy of only 420 keV and a very small

probability of escape. Moreover, for the thin convertor, the
probability of photon absorption in In is very small, say P_{In} (t < 0.5
g/cm^2) < 5%. It is independent of neutrino energy, and photoabsorption
in plastic can be neglected.

c) D_e (E_ν, t, d) is the probability of the prompt electron escape from the
converter. It is a very sensitive function of both the converter and
the plastic thickness. Furthermore, it depends on the neutrino energy.
For example, it is rather small, about 10% for low energy Be-7
neutrinos and close to 100% for very energetic B-8 neutrino.

The estimated quantum detection efficiency, QDE, permits calculation of the
mass of detector necessary to detect 0.2 neutrinos/day (see Tab. 4). It can be
seen that assuming some improvement in γ-radioactivity, a mass of indium should
be ca. 20 tons and one needs big liquid Xenon chambers. One expects detection
of 0.3 neutrinos/day, of which 67%, 21%, and 12% are due to Be-7, pep and B-12
cycles. We should like to point out that this is not the optimal design of the
detector. By improvement of spatial and energy resolution of liquid Xe MWPC,
and by fine tuning of the parameters of the experimental set-up (coincidence
time, filling factor m and thickness of In converter) the improvement of count
rate to 0.5 neutrinos/day is expected.

The above discussed results of ref. 11 have recently become of considerable
interest. In 1984, it was believed that combined results of Cl and Ga
experiments will conclusively resolve the "solar neutrino" problem. Today,
the situation is very different. The Super-Kamikande results seem to confirm
paucity of B-8 solar neutrinos (see this proceeding). On the other hand,
Ga-projects have been funded and will in a few years give results about pp solar

neutrinos. However, if the hypothesis of resonant oscillation of the neutrino is true, we need information about pp, Be-7 and B-8 neutrinos.

The advantages of detecting Be-7 neutrinos are:
- possibility to decide between resonant oscillation and other reasons for solar neutrino paucity;
- possibility of measuring the temperature of solar interior.

Thus the proposed variant of In-115 inverse beta decay is of outmost interest: it provides simple, good signal/background ratio detector of Be-7 neutrinos. Last but not least, there is rapid progress in development of noble liquid TPC, and thousand tonn detectors are being built (Iccarus project).

3. Detecting Supernovae Neutrinos

Theorists have long believed that neutrino, not photon, emission dominates the last phases of the evolution of massive stars ($M \geq 8\ M_e$). After carbon burning, the thermonuclear energy (generated at the highest temperatures and densities of the stellar core) is carried away by the weakly coupled neutrinos that can stream unimpeded out through the massive overburden. The collapse and supernova phase is thought to be accompanied by the most energetic neutrino burst of all as a neutron star (or black hole) is formed.

Kamiokande II[12] and the IMB[13] collaborations have reported the detection, at ostensibly the same time, of the neutrino burst from the supernova SN1987a in the Large Magellanic Cloud (LMC). The 8 neutrino events spread over 5.6 seconds in the IMB detector and the 11 neutrino events spread over 12.5 seconds in the Kamioka detector are a true milestone in neutrino astronomy.

This has two important sequences:
- the theory of supernovae explosions seems to be, at least qualitatively, confirmed[14];
- limits on the electron neutrino mass can be derived[15].

The standard model predicts that neutrinos of all species (ν_e, $\bar{\nu}_e$, ν_μ, $\bar{\nu}_\mu$, ν_τ, $\bar{\nu}_\tau$), <u>not</u> just ν_e's, carry away the neutron star's binding energy in roughly equal amounts in, <u>not</u> milliseconds, but <u>seconds</u> as these neutrinos <u>diffuse</u> out of the hot, opaque protoneutron star. This time scale is set by the opacity of dense matter. The initial emission is dominated by the cooling and neutronization of the shocked outer core. This emission may be enhanced by convective transport. The early phase lasts no more than half a second and blends into the long-term phase of diffusion from the inner core. As the neutrinos escape, they downscatter in energy. The average energy of the ν_e's and $\bar{\nu}_e$'s should be 10-20 MeV and that of the ν_μ's, $\bar{\nu}_\mu$'s, ν_τ's and $\bar{\nu}_\tau$'s should be 15-25 MeV.

It should be expected that the $\bar{\nu}_e$'s, by their large absorption cross section on protons, will dominate the signal in water Cherenkov detector, even though the ratios of the total energy emitted are approximately $\nu_e:\bar{\nu}_e:\nu_\mu$ = 2:40:1. Since photons cannot escape from the core during this phase, neutrinos are the only diagnostic of the event.

For an analysis of the Kamiokande and IMB data and a comparison to the standard model described above, I refer the interested read to the recent papers[14]. More specifically, the total energy was estimated to be $f_\nu = 6 \pm 3 \times 10^{52}$ ergs and the energy spectrum consistant with $kT \approx 5$ MeV. For limits on neutrino mass, the limits from 6eV to about 20eV were quoted. However, much stronger limits can be obtained (see e.g. A. Dar, this conference) if one can be sure that the two initial events of Kamikande data are really electron neutrinos. Unfortunately, the water Cerenkov detectors don't permit clean identification of neutrino flavors.

This brings the possibility of using In-115 detector for detection of local supernovae. The advantage of the In-115 based detector are:
- sensitivity to only one type of neutrinos, i.e. neutrinos produced in neutronization process,
- cross-section orders of magnitude higher than for νe scattering;
- excellent signal/noise obtained due to coincidence of the prompt electron and the 500 keV photon;
- fast, microsecond timing.

However, to be useful in galactic supernovae search, the In-115 detector should feature a mass of at least 100 tonn and good quantum detection efficiency. We are currently performing the numerical calculations[15], which suggest that such detector may be built.

REFERENCES

1. J. N. Bahcall, R. Davis, "An account of the development of the solar neutrino problem", Essays in Nuclear Astrophysics, chapter 12, Cambridge University Press 1981.

   J. N. Bahcall et al., Rev. Mod. Physics, Vol. 54 Vol. 54 No. 3 July 1982
   E. Schatzman, "The internal structure of the sun and the solar neutrinos problem", in 4th Moriond Workshop, "Massive Neutrinos in Astrophysics...", Ed. Frontiers, Paris 1984.

   J. N. Bahcall, R. Ulrich, "Solar Models, Neutrino Experiments and Helioseismology," Sept. 87 preprint IASSNS-AST87/1 (submitted to Review of Modern Physics).

2. R. Davis et al., "Solar neutrino experiment...", Massive Neutrino Mini-Conference, Telemark Wisconsin 1980.

   Ad Russian projects, see Izviesta ANCCCP, Volume 51, No. 6, p. 1231-1235, 1987 (in Russian).

3. For Ga-71 and other no-Cl experiments see: W. Hampel, "Low Energy Neutrinos in Astrophysics", Massive Neutrino Mini-Conference, Telemark Wisconsin 1980.

4. B. Pontecorvo, Sov. Phys. JETP $\underline{26}$, 984 (1968)
   V. Gribov, B. Pontecorvo, Phys. Lett. $\underline{28B}$, 493 (1969)
   J. N. Bahcall, et al., Phys. Rev. Lett. $\underline{28}$, 316 (1972)
   J. N. Bahcall, et al., Phys. Rev. Lett. $\underline{45}$, 945 (1980)
   Ad. MSW effect
   L. Wolfenstein, Phys. Rev. Lett. $\underline{56}$, 1305 (1986)
   S. P. Mikheyev, A. Yu. Smirnov, Nuovo Cim. $\underline{9c}$, 17 (1986)
   H. Bethe, Phys. Rev. Lett. $\underline{56}$, 1305 (1986)
   S. P. Rosen, J. M. Gelb, Phys. Rev. D34, 969 (1986)
   S. J. Parke, Phys. Rev. Letters 57, 12575 (1986)
   T. K. Kuo, J. Pantaleono, Phys. Rev. Lett. 57, 1805 (1986)

   for review see,
   K. Whisant in "Neutrino Masses and Neutrino Astrophysics, Telemark IV, 1987 ed. V. Berger et al.

5. R. S. Raghavan, Phys. Rev. Lett. $\underline{37}$, 259 (1976)
   M. Deutsch et al, DOE proposal, May 17, 1979
   R. S. Raghavan, "Direct detection and spectroscopy of solar neutrinos using indium", Massive-Neutrinos Mini-Conference, Telemark Wisconsin 1980

6. M. Bourdinand et al., preprint DPhPE, Saclay, January 1984
   ibid,   preprint PPhPE, Sacay, March 1984
   M. Spiro et al., "A possible solar neutrino detector based on indium reaction and using well known techniques", Proc. 4th Moriond Workshop, "Massive neutrinos...", p. 311, Ed. Frontieres, Paris 1984

7. A. K. Drukier, "Use of Indium superheated superconducting colloid as solar neutrino detector", unpublished 1979;
G. Waysand, "Detection of low energy solar neutrinos by chambers of metastable superconducting granules", Proc. 4th Moriond Workshop, "Massive neutrinos...", p. 319, Ed. Frontieres, Paris 1984.

8. A. K. Drukier, L. Stodolsky, "Principles and applications of a neutral current detector for neutrinos physics and astronomy", preprint MPI-PAE/PTh 36/82, Phys. Rev., D30, 2295-2309, 1984.

9. R. D. Evans, "The atomic nucleus", p. 625, McGraw-Hill Co., NY 1955 "Tables of Fermi functions" in "Beta- and Gamma-Ray spectroscopy", ed. K. Siegbahn, North Holland 1955.

10. N. Booth, private information.

11. A. K. Drukier, R. Nest, NIM A239, 605-622 (1985)

12. See e.g. E. W. Baier et Telemark IV Workshop, 1987 ed. V. Berger et al.

13. See e.g. T. Haines, ibid.

14. See e.g. A. Burrows, ibid.

15. See Proc. of Workshop "Supervnovae 1987A...", George Mason University, October 1987.

    Also,
    J. Bahcall et al., Nature 327, 682 (1987)
    J. Bahcall et al., preprint IASSNS-AST 87/8
    Abbott et al., BU preprint, submitted to Phys. Lett.
    A. Dar et al., Technion preprint.

16. A. K. Drukier, T. Girard, J. Mengel. In preparation.

TABLE 1: The expected capture rate for the standard solar model (SNU)

Target	p-p	^7Be	pep	^8B	total#
^{37}Cl	-	1.02	0.23	6.05	7.6
^{71}Ga	67.2	28.5	2.4	1.7	106.
^{115}In	532.	125.	9.6	5.0	700.
Al*	1.8×10^4	4.5×10^3	540.	652.	2.3×10^4
Pb*	-	4.7×10^4	5.7×10^3	6.8×10^3	5.9×10^4

* using coherent scattering on nuclei.

TABLE 2: Mass and expected quantum efficiency

Target	SNU	QDE	Mass[tons]	other mass [tons]	S/B
37 Cl	7.6	90 %	208.	–	> 3
71 Ga	106.	60 %	43.	–	> 3
115 In *	700.	40 %	1.6	100	5 - 10
115 In **	700.	20 %	3.2	2,000	1 - 5
115 In ***	140.	45 %	7.8	200	> 10
Al ****	1.8×10^4	10%	0.033	1.	> 10

* SSC colloid
** Saclay proposal
*** this paper
**** using coherent scattering on nuclei

TABLE 3: Range g/cm^2 of the prompt electrons*

Cycle	E [MeV]	E_e [MeV]	R [g/cm^2]
β	*	.490	0.159
^{7}Be	.870	.750	0.284
pep	1.44	1.32	0.577
^{8}B	*	8.0	4.13

* For β and B^8 spectrum is continuous. Quoted are E_{max} in the first and average energy in the second case.

TABLE 4: Mass of ^{115}In detector*

Cycle	SNU	QDE	Mass [tons]
^{7}Be	125	9 %	40
		18 %	20
pep	9.6	43 %	86
		60 %	62
^{8}B	5.0	80 %	112

* For 0.2 counts/day for ^{7}Be, pep and ^{8}B neutrinos, respectively. For MWPC of 10 m^2 surface, with In converter (sheet!) with thickness of 0.16 [g/cm^2]. For Be-7 and pep, the two line are for suppression factors of 10^{-4} and 10^{-3} respectively.

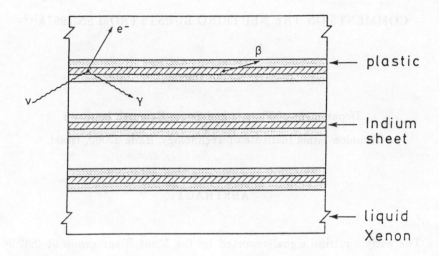

Fig. 1

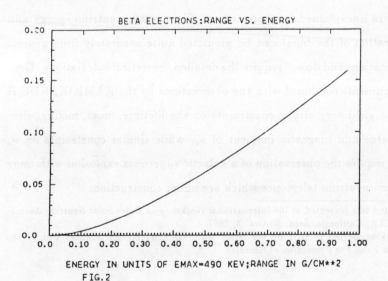

BETA ELECTRONS: RANGE VS. ENERGY

ENERGY IN UNITS OF EMAX=490 KEV; RANGE IN G/CM**2

FIG. 2

COMMENTS ON THE NEUTRINO BURSTS FROM SN 1987A[1,2]

Arnon Dar

Department of Physics and Space Research Institute
Technion-Israel Institute of Technology, Haifa 32000, Israel.

ABSTRACT

The early neutrino signal reported by the Mont Blanc group at 2:52:36 UT on February 23, 1987 could not have been produced by a neutrino burst from SN 1987A. The second neutrino burst that was reported by the KAMIOKANDE II, IMB and Baksan groups is consistent with the general theoretical picture of supernovae explosions but has definite features which are unexplained by that picture. The average neutrino energy and the duration of the burst can be predicted quite accurately from general considerations and do not require the detailed theoretical calculations. General arguments combined with the observations by the KAMIOKANDE II detector yield very strong constraints on the lifetime, mass, mixing, electric charge and magnetic moment of ν_e, while similar constraints on ν_μ and ν_τ require the observation of a galactic supernova explosion with more advanced neutrino telescopes which are under construction.

---

[1] Invited talk presented at the International Workshop on Extra Solar Neutrino Astronomy, UCLA, California, Sept. 30-Oct. 2, 1987.
[2] Work supported in part by the Israel National Academy for Sciences and by the Technion Fund For Promotion of Research.

1 DID SN 1987A BANG TWICE?

The first optical record of SN 1987A (McNaught 1987), occurred at 10:37:55 UT on February 23, 1987. The Mont Blanc scintillator detector recorded (Aglietta et al., 1987) a burst of 5 neutrino events within 7 seconds at 2:52:36 UT. About 5 hours later a second neutrino burst was detected simultaneously by the KAMIOKANDE II (Hirata et al. 1987 at 7:35:35 ± 1 minute UT) and IMB (Bionta et al. 1987 at 7:35:41:37 UT) water Cherenkov detectors and by the Baksan scintillator detector (Pomansky 1987). A few authors have consequently suggested that SN 1987A went through a two stage collapse: The first neutrino burst was generated by core collapse into a hot protoneutron star. The second burst was produced when the protoneutron star recollapsed into a black hole after it cooled and got rid of excessive angular momentum via neutrino and/or gravitational waves emission during the 4h 43m time interval (see e.g. De Rujula, 1987).

However, if the Mont Blanc 90 tonnes scintillator detector with a detection efficiency $\eta_{MB}(E_e)$ observed 5 $\bar{\nu}_e p \to e^+ n$ events with e^+ energies $E_i \pm \Delta E_i$, i = 1, ... 5, then the same $\bar{\nu}_e$ flux should have generated $17\Sigma\left[\eta_{KII}(E_i \pm \Delta E_i)/\eta_{MB}(E_i \pm \Delta E_i)\right] = 36 \pm 12$ events in the 2140 tonnes KAMIOKANDE II (KII) water detector, which contains 17 times more "free" protons and has a detection efficiency $\eta_{KII}(E_e)$. KII has not detected any event during the 7 seconds of the Mont Blanc signal. Thus it must be concluded that unless the Mont Blanc events had lower energies than estimated these events could not have been produced by a $\bar{\nu}_e$ burst from SN 1987A. Moreover, the neutrino luminosity which was required to generate the Mont Blanc signal implies that the binding energy released by SN 1987A is larger by more than an order of magnitude than the maximum binding energy released in the formation of a neutron star:

The binding energy released by neutrino emission from SN 1987A is given by

$$BE \cong 6 \times 4\pi D^2 \int E(dN_{\bar{\nu}_e}/dE)dE \,, \qquad (1)$$

where D is the distance to SN 1987A, 6 is the number of known flavours of light neutrinos and $dN_{\bar{\nu}_e}/dE$ is the energy distribution of the $\bar{\nu}_e$'s that reached the detector. It is related to the energy distribution of the $\bar{\nu}_e p \to e^+ n$ events in the detector through

$$dn/dE_e = \eta \sigma N_p (dN_{\bar{\nu}_e}/dE_\nu) \,, \qquad (2)$$

where $E_e \cong E_\nu - m_n + m_p = E_\nu - 1.3$ MeV, N_p is the number of "free" protons in the detector, and $\sigma \cong 8.9 \times 10^{-44}(E_e/MeV)^2 cm^2$. In order to determine the $\bar{\nu}_e$ flux that presumably have generated the Mont Blanc events we have assumed that it can be well represented by a Fermi-Dirac distribution with a zero chemical potential, $dN_\nu/dE \cong AE^2/[1 + exp(E/T)]$. Using Eq. (2) we determined dN_ν/dE by a maximum likelihood procedure from the Mont Blanc results and from the fact that both KII and IMB detectors have seen no events during the Mont Blanc signal. We then used it in Eq. (1) to calculate the total binding energy released by SN-1987A at a distance of D ~ 50 kpc in a neutrino burst that generated the Mont Blanc events. We obtained

$$2M_\odot c^2 \simeq 5 \times 10^{54} ergs \leq BE \leq 10^{55} ergs \simeq 4M_\odot c^2 \,. \qquad (3)$$

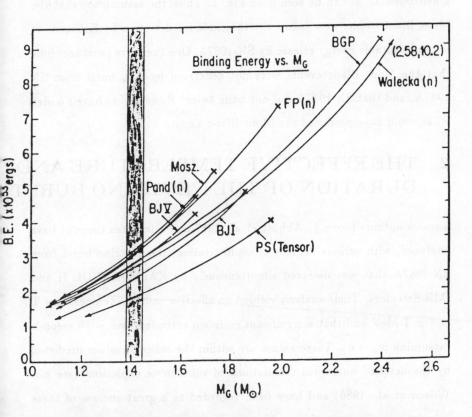

Figure 1. The binding energy versus gravitational mass for various neutron star equations of state as compiled by A. Burrows (this proceedings). The black strip indicates the range of the measured masses of neutron stars in binary systems.

This binding energy is larger by more than an order of magnitude than the estimated maximum binding energy that can be released in the formation of a neutron star, as can be seen from Fig. 1. Thus, the assumption that the Mont Blanc events were below the threshold for detection by KII leads to an unacceptable energy release by SN-1987A. One therefore must conclude that the Mont Blanc events were not generated by a $\bar{\nu}_e$ burst from SN 1987A, and that SN 1987A did not bang twice! However, we have no idea what could have produced the Mont Blanc signal.

2 THE EFFECTIVE TEMPERATURE AND DURATION OF THE NEUTRINO BURST

Various authors (see e.g. Abbott et al. 1987 and references therein) have analyzed, with various degrees of sophistication, the neutrino burst from SN-1987A that was detected simultaneously by KAMIOKANDE II and IMB detectors. Their analysis yielded an effective neutrino temperature T $\sim 4 \pm 1$ MeV and that a significant emission extended over ~ 10 seconds (assuming $m_\nu \sim 0$). These values are within the range of values predicted by the detailed numerical calculations of supernovae explosions (see e.g. Wilson et al. 1986) and have been regarded as a great success of these calculations. However, these values are expected on quite general grounds:

Due to their short mean free path in the protoneutron star the neutrinos diffuse to the surface of the protoneutron star and escape outside it when their mean free path λ_ν is of the order of the scale height,

$$\lambda_\nu \sim \rho/(\partial \rho/\partial r) \sim \alpha R \qquad (4)$$

where R is the radius of the protoneutron star and $\alpha < 1$. The water Cherenkov detectors are sensitive mainly to $\bar{\nu}_e$'s whose mean free path in the protoneutron star is given by

$$\lambda_{\bar{\nu}_e} \cong 1/\sigma N_A Y_p \rho , \qquad (5)$$

where $\sigma = \sigma(\bar{\nu}_e p \to e^+ n) \sim \sigma_o E_\nu^2$, N_A is the Avogadro number, and Y_p is the fractional proton number near the neutrinosphere from where the $\bar{\nu}_e$'s are emitted. If these $\bar{\nu}_e$'s have an approximate Boltzman distribution then their average energy is related to the temperature at the neutrinosphere via $< E_\nu^2 > \sim 12T^2$, i.e.

$$\bar{\sigma} = 12\bar{\sigma}_o T^2 \sim 12 \times 7.5 \times 10^{-44}(T/MeV)^2 \ cm^2 . \qquad (6)$$

The density ρ in the collapsing core is related to the local temperature through the equation of state (see e.g. Shapiro and Teukolsky 1983)

$$\rho \sim \rho_o T^3 \sim 1.13 \times 10^{10}(T/MeV)^3 \ g/cm^3 . \qquad (7)$$

By combining Eqs. 5-7 we finally obtain

$$T \sim \left[\frac{m_p}{12 Y_p \rho_o \alpha \bar{\sigma}_o R} \right]^{1/5} . \qquad (8)$$

Thus the temperature T depends very weakly on the precise values of Y_p, R and α. Significant neutrino emission takes place when R approaches the radius of a neutron star, which from observations is known to be $R \sim (15\pm 5)$ km. During core collapse Y_p changes from $Y_p \sim 0.45$ (iron core) to $Y_p \sim 0.04$ (cold neutron star). If we use the typical values $R \sim (? 5\pm 10)$

km, $0.1 < \alpha < 1$ and $0.04 < Y_p < 0.45$ we obtain that $T \sim 4 \pm 1$ MeV! consistent with the observation of KII and IMB.

The duration of the $\bar{\nu}_e$ burst can be estimated as follows: The neutrino luminosity of the protoneutron star is given by

$$L_\nu \sim 4\pi R_\nu^2 \sigma_\nu T^4 , \qquad (9)$$

where $\sigma_\nu = (7/16) \times 5.67 \times 10^5$ erg/deg^4 is the Stefan-Boltzman constant for left handed neutrinos, R_ν is the radius of the "neutrinosphere" and T is the temperature there. The total binding energy of a neutron star which is estimated to be (see Fig. 1) $W \sim (2.5 \pm 1) \times 10^{53}$ ergs, and which is radiated away by neutrinos is therefore given by

$$W = \Sigma_\nu \int L_\nu dt \sim N_f \bar{L}_\nu \Delta t , \qquad (10)$$

where the sum extends over $N_f = 6$ light $(m_\nu \leq T)$ neutrino flavours and Δt is the effective emission time. For $R_\nu \sim (2\,5 \pm 10)$km and $T \sim (4 \pm 1)$ MeV Eqs. (9) and (10) yield $\Delta t \sim 5\text{-}25$ seconds, consistent with the observation.

3 UNEXPECTED FEATURES OF THE NEUTRINO SIGNALS FROM SN 1987A

Many authors have pointed out that the time distribution of the KII events and the angular distribution of both the KII events and the IMB events were not expected and are quite puzzling. Other authors have dismissed them as statistical fluctuations. We believe it is scientifically irresponsible to ignore them. Instead one should explore the physical information they may store on supernovae explosions and on neutrino properties although the validation of the conclusions will have to wait until the detection of a neutrino burst from a much closer (galactic) supernova explosion.

The neutrino signals in the water Cherenkov detectors are expected to rise mainly from e^+ recoils which are produced by the reaction $\bar{\nu}_e p \to e^+ n$ with a nearly isotropic distribution with respect to the incident $\bar{\nu}_e$'s and with a cross section $\sigma(\bar{\nu}_e p \to e^+ n) \sim 8.9 \times 10^{-44}(E_e/MeV)^2 cm^2$, where $E_e \cong E_\nu - m_n + m_p = E_\nu$ -1.3 MeV. The reaction $\nu_e e \to \nu_e e$ produces electrons recoiling with $< E_e > \sim E_\nu/2$ along the direction of the incident ν_e's, but since $\sigma(\nu_e e \to \nu_e e) \sim 0.95 \times 10^{-44}(E_\nu/MeV) cm^2$ the expected number of such events is about a factor 30 smaller, assuming equal fluxes of $\bar{\nu}_e$'s and ν_e's, and T $\sim$ 4 MeV. However, the KII events contain too many events which are pointing along the direction from SN 1987A, and both the KII and IMB events show a strong energy-angle correlation, i.e. practically all the IMB and KII events well above detection threshold point in the forward hemisphere, as can be seen from Table I where we have re-ordered the KII and IMB events according to their visible energy instead of their time sequence.

In the first two events in the KII signal the e's pointed along the direction from SN1987A within experimental error ($18° \pm 18°$, $15 \pm 27°$). The chance probability that the two events were $\bar{\nu}_e p \to e^+ n$ events that accidentally produced e^+'s within a cone of 40° along the direction from SN 1987A is about 1%! These events could have neither been produced by thermal ν_e's (through the reaction $\nu_e e \to \nu_e e$) because they should have then been

Table I
Energy-Angle Correlation?

E(MeV)	$\theta°$	
35.4 ± 8.0	32 ± 16	
21.0 ± 4.2	30 ± 18	
20.0 ± 2.9	18 ± 18	
19.8 ± 3.2	38 ± 22	KII
13.8 ± 3.2	15 ± 27	
13.0 ± 2.6	40 ± 26	
12.8 ± 2.9	135 ± 23	
9.2 ± 2.7	70 ± 30	
8.9 ± 1.9	91 ± 39	
8.6 ± 2.7	122 ± 30	
7.5 ± 2.0	108 ± 32	
6.3 ± 1.7	68 ± 77	
40 ± 10	56 ± 15	
38 ± 9.5	74 ± 15	
37 ± 9.3	52 ± 15	
37 ± 9.3	52 ± 15	IMB
35 ± 8.8	63 ± 15	
29 ± 7.3	40 ± 15	
24 ± 6	102 ± 15	
20 ± 5	39 ± 15	

accompanied by about 60 $\bar{\nu}_e p \to n e^+$ events within the first 0.11 seconds between events no. 1 and no. 2. These two events however could have been produced by a neutronization burst ($e^- p \to \nu_e n$) which is much stronger (practically complete neutronization of $\sim 2 M_\odot$ iron core) and much more energetic ($E_\nu \sim 40$ MeV) than that expected in standard supernovae theory. (In standard supernova theory the neutronization burst is suppressed by Pauli blocking because of neutrino trapping in core-collapse). Thus, if the first two events were $\nu_e e \to \nu_e e$ events which were produced by ν_e's from the neutronization burst that preceded the thermal burst, they imply that neutrinos are not trapped efficiently in the collapsing core, perhaps due to

(a) Strong Convection in Collapsing Cores,

and/or

(b) Angular Momentum Effects which Modify Core-Collapse,

and/or

(c) Neutrino Magnetic Moment which Induces Helicity Flip. (Dar 1987).

Because of lepton number conservation no more than 2-3 events of the joint sample of KII and IMB events could have been due to $\nu_e e \to \nu_e e$ interactions, the rest being due to $\bar{\nu}_e p \to e^+ n$ interactions which produce nearly isotropic e recoils. Therefore the energy-angle correlation is hard to understand. Perhaps it tells us that the reaction $\bar{\nu}_e p \to e^+ n$ is not as isotropic as calculated from the static quark model (the static quark model predicts e.g., that $g_A/g_V = 5/3$ while the experimental value $g_A/g_V = 1.27$ can be predicted only from a more precise picture of the nucleons such as provided by the Cloudy Bag Model) or that the cross section for $\nu_e O^{16} \to e F^{17}$ is larger than expected and peaks in the forward hemisphere. Both cross sections should be measured by exposing water Cherenkov detectors to ν_e's from μ_{e3} decays in meson factories. (The energy spectrum of ν_e's and $\bar{\nu}_e$'s from μ_{e3} decays of stopped muons from π decays nicely overlap with the energy spectrum of neutrios from supernovae explosions).

We also note that, in principle, it is possible that some of the events in KII and IMB detectors were due to $e^{\pm}$'s from μ_{e3} decays of low energy muons that either were produced in the detectors or in the rock near it (D. Kielczewska, private communication), but it is hard to conceive a mechanism by which a supernova explosion generates such large fluxes of low energy muons in the detectors or in the rock near it.

4 NEUTRINO PROPERTIES FROM OBSERVATION OF SN 1987A

Table II compares constraints on the properties of the neutrinos that were obtained from experiments with terrestrial neutrino sources and constraints that were derived by us and by other authors from observations of the neutrino burst from SN 1987A. Question marks indicate that the conclusions are based on assumptions that may be disputed.

4.1 Neutrino Life Time

The idea of unstable neutrino was proposed many years ago by Bahcall et al. (1972) and by Pakvasa and Tennakone (1972) as a solution to the

Table II
Comparison Between Electron Neutrino Properties As Determined From Terrestrial Neutrinos And From SN 1987A Neutrinos

Property	Terrestrial Experiments	SN 1987 A
Life Time ($\gamma\tau_\nu$ in sec)	Atmospheric ν's: $> 4 \times 10^{-2}$ Solar ν's : $\geq 5 \times 10^2$	$> 5 \times 10^{12}$ (if no mixing)
Mass (m_{ν_e} in eV/c^2)	< 18 (Zurich) $3\pm$ (Lubimov)	$< 3.4 \pm 1$? < 15 ?
Mixing With ν_L ($m_{\nu_L} < MeV$)	$\Delta m^2 \geq 0.1 eV^2$ $\sin^2 2\Theta > 0.1$ $\Big\}$ excluded	Practically Excluded? (See Fig. 2)
Mixing With ν_h ($m_{\nu_h} > MeV$)	Left Narrow Window	Ruled Out if $1 < m_{\nu_h} < 100$ MeV
Electric Charge (q/e)	$< 10^{-13}$	$< 2 \times 10^{-17}$
Magnetic Moment (μ/μ_B)	$< 10^{-9}$	$< 10^{-13}$ $< 2 \times 10^{-15}$?
ν Flavours	≤ 4 (Cosmology)	$\leq 4 \pm 1$

solar neutrino problem (Bahcall et al. 1982) and has been recently revived (Bahcall et al. 1986) in the light of theoretical models for ν-instability (Gelmini and Valle, 1984, Valle et al. 1987). The observation of $\bar{\nu}_e$'s and (?) ν_e's with E $\sim$ 15 MeV from SN 1987A (Hirata et al. 1987, Bionta et al. 1987, Aglietta et al. 1987, Pomansky 1987) has been widely perceived ruling out ν-decay as an explanation to the solar neutrino problem (Bahcall et al. 1987A). In the theoretically preferred regime of small flavour-mixing and small mass difference the observation of the ν-burst from SN 1987A indeed implies that $\gamma\tau > 5 \times 10^{12}$ sec for ν_e's. However, it has been pointed out recently (Frieman et al. 1987, Raghavan et al. 1987) that if $\bar{\nu}_e$'s and ν_e's are unstable but admixed significantly with other neutrinos which are stable, then they could have generated the observed neutrino signals from SN 1987A. (This however may require a neutrino luminosity which is too large compared with the binding energy release expected from a type II supernova like SN 1987A).

4.2 Neutrino Mass

The basic idea was discussed first by Zatsepin (1968) who pointed out that if neutrinos had a finite mass, the higher energy neutrinos from a supernova explosion would arrive before the more slowly moving, lower energy neutrinos. The difference in time of flight from SN 1987A to Earth (distance D) of two neutrinos of energies E_1 and E_2, respectively, is given by $\Delta t \cong (D/2c)(m_\nu c^2)^2(E_1^{-2} - E_2^{-2})$. A finite mass of the neutrino will cause, accordingly, particles of different energies to arrive at different times. This dispersion in arrival times can be used to extract a neutrino mass if one makes assumptions on the emission time. For instance many independent authors (see e.g. Abbott et al. 1987 and references therein) have assumed that the neutrino luminosity ($L_\nu \sim R^2 T^4$) can be fitted by a smoothly varying function of time, and from a statistical analysis deduced that $m_\nu < 15$ eV. However, some authors (e.g. Arafune and Fukugita 1987) pointed out that the first two events observed by KAMIOKANDE II are most probably $\nu_e e$ scattering events generated by electron neutrinos from the neutronization burst. (These events were both observed in the forward direction with respect to the LMC-Earth axis). A neutronization pulse should precede almost all of the neutrino emission, but it is predicted to be rather weak in the standard theory. However, the standard theory has not incorporated angular moment effects during the collapse, neither neutrino transport via convection, nor neutrino magnetic moment effects (if the neutrino has a magnetic moment), etc., which may lead to an enhanced neutronization burst (limited only by lepton number conservation). The 4 eV upper limit on electron neutrino mass was obtained by Dar and Dado 1987a by requiring that events 3, 4, 5, in the KAMIOKANDE detector were generated by $\bar{\nu}_e$'s (as evident from the large angles of the recoiling electrons) which were emitted in the thermal burst <u>after</u> the ν_e's from the neutronization burst which generated events 1 and 2.

The validity of the various assumptions used by different authors in extracting limits on m_{ν_e} will be tested only when much larger neutrino signals from galactic supernovae explosions will be detected. Moreover, larger signals from galactic supernovae explosions and the different sensitivity of some future underground neutrino detectors to different neutrino flavours (e.g. D_2O detectors) will yield very significant bounds on the ν_μ and ν_τ masses, $m_{\nu_\mu}, m_{\nu_\tau} \leq 10$ eV compared with present bounds which are $m_{\nu_\mu} <$ 250 KeV and $m_{\nu_\tau} < 56$ MeV (A. Dar et al. 1988).

4.3 Mixing of ν_e With Other Generations

If the first two events which were generated by neutrinos from SN 1987A in the KAMIOKANDE II detector were induced by ν_e's from the neutronization burst then the ν_e's from the neutronization burst could not suffer flavour flip due to the MSW effect (Mikheyev and Smirnov 1986, Wolfenstein 1978) in the progenitor envelope, because their inferred number is about the maximum number allowed by lepton number conservation for a complete neutronization of an iron core with $M_c < 2M_\odot$. Since the range of densities encounterd in the progenitor envelope includes the range of densities encountered in the sun, the absence of any significant MSW conversion of ν_e's into another flavour in the progenitor envelope therefore indicates that the MSW effect is not responsible for the solar neutrino problem (Dar and Dado 1987a, Arafune et al. 1987, Notzold 1987). In fact, the absence of any significant evidence for the MSW effect in the progenitor envelope excludes such a large range of ν_e mixing parameters that practically excludes ν_e mixing with light ν's, as indicated in Fig. 2.

4.4 Mixing of ν_e With ν_i, 1 MeV$< m_{\nu_i} <$100 MeV:

Mixing of the ν_e with ν_i induces muon like decay $\nu_i \to \nu_e e^+ e^-$, if $m_{\nu_i} > 2m_e$. Experiments searching for such decays of ν_τ's and of higher neutrino genera-

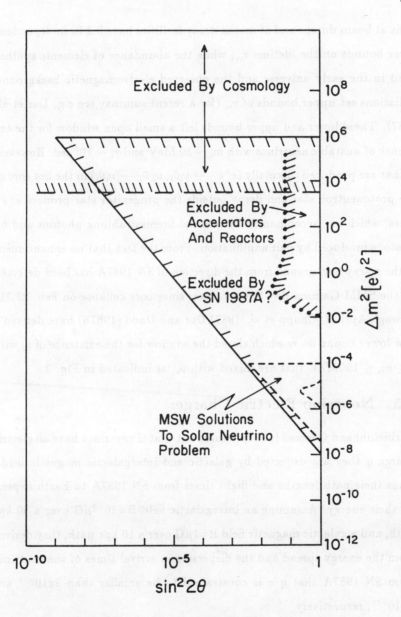

Fig. 2: Values of the mixing parameters of electron neutrinos which are excluded by accelerators and reactor experiments, by cosmological arguments and by the neutrino observation of SN 1987A (?)

tions at beam dumps and at meson decay facilities have led to m-dependent lower bounds on the lifetime τ_{ν_i}, while the abundance of elements synthesized in the early universe and the observed electromagnetic background radiations set upper bounds of τ_{ν_i} (for a recent summay see e.g. Dar et al. 1987). These lower and upper bounds left a small open window for the existence of unstable neutrinos with $m_i \sim 50$ MeV and $\tau_i \sim 10^3$ sec. However, ν_i that are produced thermally ($e^+e^- \to \nu_i\bar{\nu}_i$, $\nu_L\bar{\nu}_L \to \nu_i\bar{\nu}_i$) in the hot core of the protoneutron star and decay outside the progenitor star produce e^+e^- pairs, which are accompanied by internal bremsstrahlung photons and by photons produced by pair annihilation. From the fact that no enhancement of the γ-ray background from the direction of SN 1987A has been detected by the SMM Gamma Ray Spectrometer since core collapse on Feb. 23,316 through April 9 (Chupp et al. 1987) Dar and Dado (1987b) have derived a new lower bound on τ_i which closed the window for the existence of ν_i with $1 \leq m_i \leq 100 MeV$ that are mixed with ν_e, as indicated in Fig. 3.

4.5 Neutrino Electric Charge:

Barbiellini and Cocconi (1987) pointed out that if neutrinos have an electric charge q they are deflected by galactic and intergalactic magnetic fields. Thus their path lengths and flight times from SN 1987A to Earth depend on their energy. Assuming an intergalactic field $B=10^{-3}\mu G$ over a 50 kpc path, and a galactic magnetic field $B=1\mu G$ over a 10 kpc path, they derived from the energy spread and the dispersion in arrival times of the neutrinos from SN 1987A that q/e is constrained to be smaller than 2×10^{-15} and 2×10^{-17}, respectively.

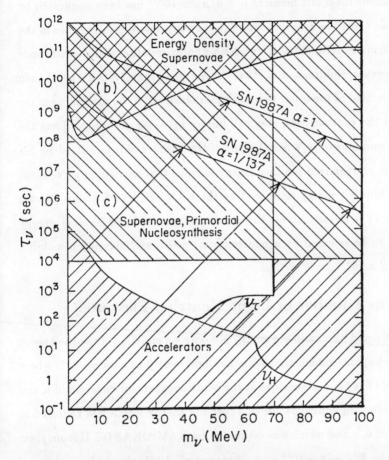

Fig. 3. Constrains on the lifetime of heavy neutrinos as function of their mass, reproduced from Dar and Dado 1987 and references therein. The shaded areas are excluded by accelerator experiments and by previous astrophysical and cosmological observations. The new lower bounds obtained from SN 1987A are indicated by the arrows.

4.6 Neutrino Magnetic Moment:

An anomalous magnetic moment $\eta = \mu/\mu_B \simeq 10^{-10}$ has been suggested by Okun et al. (1986) as an explanation to the suppressed counting rate in the solar neutrino experiment and its possible anticorrelation with the sun spot cycle. However, Goldman et al.(1987) have argued that if $\eta > 2 \times 10^{-15}$ then a magnetic field of the order $B \sim 10^{12}$ gauss believed to be present near the surface of the protoneutron star will induce $\nu_L \leftrightarrow \nu_R$ precesion. Thus the observed numbers of ν_L's is only half of the number of ν's emitted by SN 1987A, i.e. the inferred binding energy is twice that carried by ν_L's. However, the binding energy estimated by the KAMIOKANDE group (Hirata et al. 1987) is already about the maximum gravitational binding energy expected to be released by a neutron star, which excludes the possibility that $\eta > 2 \times 10^{-15}$.

4.7 The Number of Light Neutrino Flavours:

SN 1987A could have radiated energy not only via emission of the the known neutrinos but also via emissions of other light neutrinos with $m_i < T_i$ where $T_i \sim 10$ MeV is the temperature at the neutrinosphere for emission of μ and τ like interacting neutrinos. In this case the energy which was radiated in the form of $\bar{\nu}_e$'s and which was estimated by KAMIOKANDE II from their events to be $W_{\bar{\nu}_e} \sim 8 \times 10^{52}$ ergs (Hirata et al. 1987) should be multiplied by N_f, the number of light neutrino flavours, to obtain the total binding energy that SN 1987A has released. If this total binding energy is smaller than 4×10^{53} ergs, as expected from the best available equation of states of nuclear matter, then $N_f < 5$.

REFERENCES

Abbott, et al. (1987), Preprint, BUHEP-87-24.

Aglietta M., et al. (1987), Europhysics Lett. 3, 1315.

Arafune, J. and M. Fukugita, (1987), Phys. Rev. Lett. 59, 367.

Arafune, J. et al., (1987), Phys. Rev. Lett. 59, 1864.

Bahcall, J.N. et al. (1972), Phys. Rev. Lett. 28

Bahcall, J.N. et al. (1982), Phys. Mod. Phys. 54, 767.

Bahcall, J.N. et al. (1986), Phys. Lett. 181B., 369.

Bahcall, J.N. et al. (1987a), Nature 326, 125.

Barbiellini, G. and Cocconi G. (1987), Nature 329, 21.

Bionta, R.M. et al. (1987), Phys. Rev. Lett. 58, 1494.

Chupp, E.L. et al. (1987), Phys. Rev. Lett. 58, 2146.

Dar, A. (1987), IAS Preprint, February 1987.

Dar, A. et al.(1987), Phys. Rev. Lett. 58, 3246.

Dar, A. and Dado, S. (1987a), Preprint Technion-PH-000.

Dar, A. and Dado, S. (1987b), Phys. Rev. Lett. 59, 2368.

Dar, A. et al., (1988), to be published.

De Rujula, A. (1987), CERN Preprint TH-4702/87.

Frieman, J.A. et al. (1987), Preprint SLAC-PUB 4261.

Gelmini, G. and Valle, J.W.F. (1984), Phys. Lett. 142B, 181.

Goldman, I. et al. (1987), Preprint TAUP 1543-87.

Hirata, K. et al. (1987), Phys. Rev. Lett. 58, 1490.

Mikheyev, S.P. and Smirnov A. Yu (1986), Nuovo Cimento C9, 17.

Notzold, D. (1987), Preprint MPI-PAE/PTh 09/87.

Okun L.B. et al. (1986), Preprint ITEP 82-86.

Pakvasa S. and Tennakone (1972), Phys. Rev. Lett. 28, 1415.

Pomansky, A. (1987), Proceedings of the XXIII Rencontre De Moriond (Ed. Tran Thanh Van).

Raghavan, R.S. et al. (1987), Preprint.

Shapiro, S.L. and Teukolsky, S.A. (1983), "Black Holes, White Dwarfs and Neutron Stars", p. 536 (John Wiley and Sons., Publ. Comp. New York).

Spergel, R.S. et al. (1987), Preprint.

Valle, J.W.F., et al. (1987), Phys. Lett. 131B, 83.

Wilson, J.R. et al. (1986), Ann. N.Y. Acad. Sci. 470, 267.

Wolfenstein, L. (1978), Phys. Rev. D17, 2369.

Zatsepin, G.I. (1968), JETP Lett. 8, 205.

D. THEORETICAL IDEAS AND FUTURE DETECTORS

Solar Monopoles and Terrestrial Neutrinos[*][†]

JOSHUA FRIEMAN

Stanford Linear Accelerator Center
Stanford University, Stanford, California 94305

ABSTRACT

Magnetic monopoles captured in the core of the sun may give rise to a substantial flux of energetic neutrinos by catalyzing the decay of solar hydrogen. We discuss the expected neutrino flux in underground detectors under different assumptions about solar interior conditions. Although a monopole flux as low as $F_M \sim 10^{-24}$ cm^{-2} sec^{-1} sr^{-1} could give rise to a neutrino flux above atmospheric background, due to $M\bar{M}$ annihilation this does not translate into a reliable monopole flux bound stronger than the Parker limit.

---

[*] Invited talk presented at the Extra Solar Neutrino Astronomy Conference, UCLA, Sept. 30 - Oct. 2, 1987.
[†] Work supported by the Department of Energy, contract DE-AC03-76SF00515.

Since they were first demonstrated to be generic features of grand unified theories, magnetic monopoles have been of enduring interest to theorists and experimentalists alike. Since a new generation of detectors, designed in part to search for these objects, is now being completed, it is useful to review the current limits on the monopole abundance. The strongest astrophysical bounds on the flux of superheavy monopoles come from the survival of pulsar magnetic fields,[1] $F_M \lesssim 10^{-24}$ cm^{-2} sec^{-1} sr^{-1}, and from monopole-catalyzed nucleon decay (Callan-Rubakov effect) in neutron stars,[2] $F_M \sigma_{28} \lesssim 10^{-21}$ cm^{-2} sec^{-1} sr^{-1} or $F_M \sigma_{28} \lesssim 10^{-28}$ cm^{-2} sec^{-1} sr^{-1} if main-sequence capture is included [3] [here the catalysis cross-section $\sigma_c = \sigma_{28} 10^{-28}$ cm^2]. These limits rely to some extent on untested properties of neutron stars, so it is natural to consider a better understood system which we can observe at close range, like the sun. In particular, it has been suggested that the Callan-Rubakov effect in the solar interior could give rise to an observable high-energy neutrino signal in underground detectors,[3][4] yielding a limit on the monopole flux comparable to (and more reliable than) the neutron star bounds.[4] In this talk, I review the expected neutrino signals from solar monopoles and critically examine the proposed limits.

We must first estimate the monopole population in the sun. The monopole stopping power in a stellar plasma is roughly [5] $dE/dx \simeq 10(g/g_D)^2 \rho(v_{mon}/c)$ GeV cm^{-1}, where (g/g_D) is the monopole charge in Dirac units and ρ is the density in gm cm^{-3}. From this, one can show [3] that the initial kinetic energy of the fastest monopole that can be captured in stars is $\frac{1}{2}mv_\infty^2 \lesssim 2.4 \times 10^{11}(g/g_D)^2$ GeV, approximately independent of stellar mass. Thus, a significant fraction of monopoles with the galactic virial velocity, $v_\infty \simeq 10^{-3}c$, are captured if their mass $m \lesssim 5 \times 10^{17}(g/g_D)$ GeV, which includes typical GUT-scale monopoles. A detailed calculation shows [3] that the total number captured over the lifetime of the sun is $N_\odot^{cap} = 3 \times 10^{41} F_M$, where F_M is the galactic monopole flux in cm^{-2} sec^{-1} sr^{-1}.

Once captured in the sun, monopoles sink to the solar core, where they efficiently destroy nucleons. Given a solar nucleon density n_n, the catalysis rate per monopole is just $n_n \langle \sigma_c v \rangle$. Since the typical energy released in a catalysis reaction is of order the nucleon rest mass, the resulting neutrino flux at the earth is approximately

$$F_\nu^{cat} \simeq \frac{N_\odot^{mon} n_n m_n c^2 \langle \sigma_c v \rangle f_c}{4\pi R_{es}^2 \langle E_\nu \rangle} , \qquad (1)$$

where $f_c \sim 1$ is the average number of neutrinos produced per nucleon decay, $\langle E_\nu \rangle \sim 200$ MeV is the average neutrino energy, and $R_{es} = 1.5 \times 10^{13}$ cm is the mean earth-sun distance. The catalysis cross-section is expected to scale as v^{-1}, which would yield $\langle \sigma v/c \rangle = \sigma_{28} 10^{-28}$ cm². However, Arafune and Fukugita[6] found that the cross-section on hydrogen is enhanced by a velocity-dependent factor $F(v/c) \simeq 170(10^{-3} c/v)$. Using $v/c \simeq 10^{-3}$ for nucleons in the solar interior (the monopole thermal speeds are negligible), from Eqn. (1) we find an expected neutrino flux $F_\nu^{cat} \simeq (N_\odot^{mon} \sigma_{28}/10^{17})$ cm⁻² sec⁻¹. Using the limit on the flux of high energy neutrinos from proton decay detectors, $F_\nu(E_\nu \gtrsim 200 \text{ MeV}) \lesssim 1$ cm⁻² sec⁻¹, we obtain an upper bound on the number of monopoles in the sun,

$$N_\odot^{mon} \sigma_{28} \lesssim 10^{17} . \qquad (2)$$

If we assume the number of monopoles in the sun is just the number captured, $N_\odot^{mon} = N_\odot^{cap} \simeq 10^{41} F_M$, we find the stringent "solar catalysis flux bound" $F_M \sigma_{28} \lesssim 10^{-24}$ cm⁻² sec⁻¹ sr⁻¹.

The final assumption above, while seemingly plausible, neglects the fact that monopoles and antimonopoles in the sun are likely to annihilate each other. Since they are so heavy, monopoles cluster deep in the core; the characteristic radius of the solar monopole distribution supported by thermal pressure is only $r_{TH} \simeq 100 m_{16}^{-1/2}$ cm, where the monopole mass $m_M = m_{16} 10^{16}$ GeV. For conditions in the solar center, it turns out that three-body annihilation processes dominate,[7] with cross-section $\langle \sigma_{ann} v \rangle = n_M m_M^{-1/2} \alpha^{-5} (kT)^{-9/2} \simeq 9 \times 10^{-29} m_{16}^{-1/2} n_M$ sec⁻¹. The solar monopole number approaches a fixed point where the annihilation and capture rates balance, given by $N_\odot^{mon} = N_{TH} \simeq 10^{22} m_{16}^{-1/2} F_M^{1/3} \ll N_\odot^{cap}$.

Combining this with the limit of Eqn. (2) yields a much weaker but more reliable solar catalysis bound on the monopole flux, $F_M \sigma_{28} \lesssim 10^{-15}$ cm^{-2} sec^{-1} sr^{-1} [see Fig. 1]. Note that this is comparable to the Parker bound,[8] $F_M \lesssim 10^{-16}$ cm^{-2} sec^{-1} sr^{-1}, coming from the survival of the galactic magnetic field.

If $M\bar{M}$ annihilation is so prolific in the sun, we may wonder whether annihilation rather than catalysis could generate an observable neutrino signal. The neutrino flux from monopole annihilation is obtained from Eqn. (1), with the replacements $n_n \to n_M$, $m_n \to m_M$, $\sigma_c \to \sigma_{ann}$, $f_c \to f_{ann}$, and using $N_\odot^{mon} = N_{TH}$. Assuming $f_{ann} \sim 1$, we find a neutrino flux $F_\nu^{ann} \simeq m_{16}^{3/2}(F_M/10^{-16})^{2/3}$. Again applying the detector limit on high energy neutrinos, one obtains a 'solar annihilation bound' comparable to the Parker limit.

For completeness, we should note that the occurence of $M\bar{M}$ annihilation in the sun itself depends on unknown conditions deep in the solar core. For example, a static magnetic field of strength $B \gtrsim 400$ Gauss would separate the monopole and antimonopole distributions sufficiently to prevent annihilation. Such a field would have to be primordial (no dynamo operates in the convectively stable core) and would have to have a coherence length $l \gtrsim 10^9$ cm in order to be stable against resistive decay on the solar lifetime. In addition, since magnetic flux tubes are buoyant and generally diffuse toward the stellar surface, the field must be sufficiently weak to be anchored at the center for 10^9 years. Although this is a possible mechanism for $M\bar{M}$ separation, and *might* lead to a large neutrino signal from catalysis, it is by no means highly probable. As a result, magnetic fields cannot be invoked to 'save' the more stringent solar catalysis bound.

We conclude that underground observations of the high energy neutrino flux place bounds on the solar monopole abundance which are competitive with the Parker limit. These solar bounds cannot reliably be made more restrictive. In particular, they do not preclude the direct detection of monopoles by large detectors such as MACRO. On the other hand, *if* magnetic fields separate monopoles from antimonopoles in the solar core, a monopole flux as low as $F_M \sim 10^{-24}$

cm^{-2} sec^{-1} sr^{-1} would generate a high energy neutrino flux from catalysis which should be separable from atmospheric background.

Acknowledgements:

I thank my collaborators Katherine Freese and Michael Turner.

REFERENCES

1. J. Harvey, M. Ruderman, J. Shaham, *Phys. Rev.* **D33** (1986), 2084.

2. For a review, see E. Kolb and M. Turner, *Ap.J.* **286** (1984), 702.

3. J. Frieman, K. Freese, M. Turner, Ap.J., to appear. J. Frieman, in "Santa Fe 1984, Proceedings, The Santa Fe Meeting," p. 476.

4. J. Arafune, M. Fukugita, *Phys. Lett.* **133B** (1983), 380. A. Dar and S. Rosen, Los Alamos preprint (1984).

5. S. Ahlen and G. Tarle, unpublished (1983).

6. J. Arafune and M. Fukugita, *Phys.Rev.Lett.* **50** (1983), 1901.

7. D. Dicus, D. Page, V. Teplitz, *Phys. Rev.* **D26** (1982), 1306.

8. E. Parker, *Ap.J.* **163** (1971), 225. M. Turner, E. Parker, T. Bogdan, *Phys. Rev.* **D26** (1982), 1296.

1) The solar monopole abundance as a function of the monopole flux. $N_\odot^{cap}$ is the number captured over the solar lifetime, and N_{TH} is the equilibrium solar abundance taking annihilations into account. The long horizontal line is the solar catalysis limit; the short horizontal shows the neutron star catalysis limit for comparison.

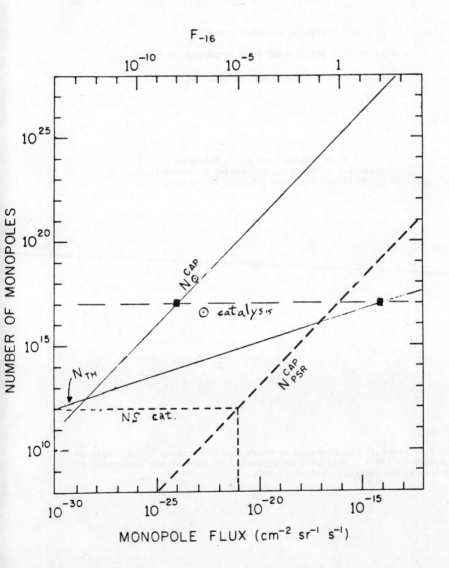

SUPERNOVA NEUTRINO DYNAMICS:

WHAT WE HAVE NOT LEARNED FROM SUPERNOVA 1987a AND

WHAT TO EXPECT FROM THE COMING GALACTIC SUPERNOVA.*

S. A. Bludman and P. J. Schinder
Department of Physics, University of Pennsylvania
Philadelphia, Pennsylvania 19104-6396

---

*Invited paper at the Workshop on Extra Sola ..stronomy (UCLA, Sept. 30 - October 1, 1987). This work is supported in part by the Department of ᵣ-ergy under contract EY-76-C-02-3071.

I. Introduction

A. The Observed Neutrino Data

We have analyzed[1] the arrival times and energies of the combined Kamiokande and IMB neutrino events for consistency and synchronicity and for information on the cooling of the nascent hot neutron star. In this analysis, we assumed only that the two detectors are observing a single astronomical happening and we ignored the data on neutrino arrival directions and the possibility of labelling the flavor of any individual neutrinos. We treated all 19 detected neutrino as $\bar{\nu}_e$, although a sample this size would be expected to contain some electron scattering events of other flavor $\overset{(-)}{\nu}_i$. If the sample were large enough and the angle measurements were precise enough, then $\overset{(-)}{\nu}_i$ scatterings could be distinguished from $\bar{\nu}_e$ statistically. Even if scattering events could be individually identified, the large uncertainties in the measured scattering angle would allow only lower limits to be placed on the energies of the incident $\overset{(-)}{\nu}_i$.

Fig. 1 (from ref 2) shows the energies and arrival times of the 11 Kamiokande (■) and 8 IMB (▲) events with their energy uncertainties of about 25%. The arrival times are measured to better than millisecond precision, but the absolute times at Kamiokande are uncertain by $\pm 1^m$. In the figure, the first Kamiokande event is plotted at the same time, t = 0, as the first IMB arrival. Two of our goals are, in fact, to determine the consistency of the two data sets and to fix the time offset of the Kamiokande clock relative to the absolute time measured at IMB.

The dashed curve connecting the 8 IMB events and the dotted curve connecting the 11 Kamiokande events[2] show that IMB is detecting generally higher energy neutrinos, Kamiokande lower energy neutrinos, but that at both detectors the neutrino energy and counting rate decreases rapidly in the first two seconds and slowly thereafter: the last two IMB events arrive 3-4 seconds after the first bunch of six, and the last three Kamiokande events arrive 7-10 seconds after the first bunch of eight. This suggests cooling of the neutrino source. Unless IMB is underestimating its sensitivity and/or Kamiokande is overestimating its sensitivity, it appears as if the lower energy Kamiokande data stream should be starting after the IMB data stream starts. The offset cannot be very large, however, because most events are bunched in the first two seconds and because the last IMB events are needed to help populate the seven second gap at Kamiokande, which by itself would have only 0.4% Poisson probability.

Finally, because Kamiokande sees some higher energy events arriving two seconds after lower energy events, the original pulse at the source must be at least two seconds in duration. Together with the failure to discriminate $\overset{(-)}{\nu}_i$ flavors, this means that, even if neutrinos of energy E_ν (Mev) have finite rest mass m_ν (eV), the dispersion in arrival times from the large Magellenic cloud, assumed to be at D = 50 kpc,

$$\Delta t = 2.5 \ (M_\nu(eV)/E_\nu(Mev))^2 \ \text{seconds},$$

so that from the SN 1987a data one cannot resolve any neutrino mass $\leq$ 10 eV. For this reason, none of the many papers on neutrino mass from SN 1987a improves very much on the laboratory upper limit $m_{\nu_e} \leq$ 20 eV. In our analysis we neglect all neutrino masses.

B. Goals

Neglecting the scant information contained in the angular distribution, the small (<10%) contamination of the $\bar{\nu}_e$ by other flavor neutrinos, any neutrino masses and any initial heating phase in the $\bar{\nu}_e$ source, our goals are to:

(1) Fit the energies and arrival times of all 11+8 neutrino events to a thermal source emitting $N_o \bar{\nu}_e$ of total energy E_o.

(2) Determine the offset of the Kamiokande arrival times with respect to the universal times recorded at IMB.

(3) Search for any other structure in the time signature of the original neutrino source. A 0.1 second dynamical period would be characteristic[3,4] of the densities $\sim 10^8$ g cm^{-3} out of which the neutrinos are expected to emerge.

Our statistical methodology comprises

(1) Maximum likelihood for the <u>best fit parameters</u> of our cooling model;

(2) Monte Carlo simulation at the detectors to determine <u>statistical uncertainties</u> in these parameters;

(3) "Kolmogorov-Smirnov by eye" comparison between our model integral spectra and number of arrivals and the data histograms, in order to determine <u>goodness of fit</u>.

We will first review models in which the source remains at constant temperature until it turns off (Sec. II), then cooling models (Sec. III). In Sec. IV we search for any periodicity in the time signal of the source. Finally, we summarize our modest conclusions from SN 1987a (Sec. V) and discuss the more promising prospects for a stellar collapse in our own Galaxy (Sec. VI).

II. Constant Temperature Models

If all the 11 Kamiokande + 8 IMB neutrino events originate in a source that is on for 12.439 seconds at temperature T_0 emitting a number N_0 of $\bar{\nu}_e$ with energy $E_0/N_0 = 3.15\, T_0$, then our maximum likelihood fit to the combined data is

$$T_0 = 3.3 \pm 0.3 \text{ MeV}, \quad N_0 = (3.4 \pm 1.0) \times 10^{57}\, \bar{\nu}_e$$

$$E_0 = (5.7 \pm 1.7) \times 10^{52} \text{ ergs}.$$

The quoted errors are one standard deviation errors derived from Monte Carlo simulations of the Kamiokande and IMB detectors using the central values determined by maximum likelihood.

If we were to exclude the last 3 Kamiokande events, we would have a higher temperature and lower fluence and energy $T_0 = 3.7$ MeV, $N_0 = 2.0 \times 10^{57}\, \bar{\nu}_e$, $E_0 = 3.7 \times 10^{52}$ ergs, in good agreement with Bahcall et al[5], who also did maximum likelihood analysis. We see, however, no objective basis for discriminating against any neutrino events, and consistently deal with the entire 11 + 8 neutrino sample.

Burrows and Lattimer[6] and Schramm[7] compare numerical moments of the actual data with equivalent moments of theoretical distributions. While the maximum likelihood function contains all of the information (gaps as well as events), the moment methods rely on the the 19 events being fair representations of the underlying distribution. While both methods should agree asymptotically, when applied to small data sets maximum likelihood is preferred[8].

Because the time does not appear explicitly, constant temperature models contain no rationale for determining the time offset between the Kamiokande and IMB data. By assuming implicitly abrupt turn-off after 12.4 seconds, constant temperature models underestimate N_o, E_o.

III. Continuous Cooling Models

Continuous cooling of the neutrino source is indicated by (1) the decrease in average energy and counting rate with time, at each detector; (2) the long (7 second) gap in the Kamiokande data stream before the last 3 events appear; (3) the end of the Kamiokande and IMB data streams when the neutrino energies fall below detection threshold.

We therefore considered[1] cooling models in which the temperature decreased continuously as a power law in time

$$T(t) = T_s (1 + at/n)^{-n} \tag{1}$$

The limiting case $n \to \infty$ is the exponential $T(t) = T_s \exp(-at)$ with constant cooling time $\tau = a^{-1}$. For any $n < \infty$, however, the cooling time

$$\tau(t) \equiv - (d\, \ln T/dt)^{-1} = a^{-1} + t/n$$

is initially short, $\tau(0) = a^{-1}$, but becomes prolonged in time; the smaller $\underline{n}$, the more protracted is the ultimate cooling. Indeed, because $N_o \propto \int T^3 dt \propto (3n-1)^{-1}$

and $E_o \propto \int T^4 dt \propto (4n-1)^{-1}$, we must have $n > 1/3$ and the time-averaged temperature $<T> = T_s(3n-1)/(4n-1) \approx 9 T_s(n-1/3)$ for $n \approx 1/3$ and $<T> = T_s(3/4)$ for exponential cooling. The time averaged neutrino Energy $E_o/N_o - 3.15<T> = 2.36 T_s$ for exponential cooling and $28 T_s(n-1/3)$ for $n \approx 1/3$.

The cooling law (1) is realized, for example, when a stellar core of fixed radius R and heat content

$$U(t) \propto \frac{4}{3} \pi R^3 T^m \qquad m = 4 - 1/n > 1$$

radiates as a black body

$$-\frac{dU}{dt} = 4\pi R^2 \sigma T^4.$$

For a semi-degenerate Fermi gas, we would expect $m = 2$ or $n = 1/2$.

This power law cooling model is the simplest extension of the earlier constant temperature models which fitted two parameters T_o, N_o. Once $\underline{n}$ is chosen, this simple cooling model fits only one additional parameter, the initial cooling rate $\underline{a}$. By adjusting more parameters, we might describe finer temporal structure in the neutrino signal. By letting the radius R decrease initially with time, we might model a Kelvin-Helmholtz heating phase preceding the black body cooling. While a cooling model is needed, the sparse data does not justify introducing any more parameters beyond $\underline{a}$.

Our best fits to the 11 Kamiokande + 8 IMB events are presented in Table 1 and plotted in Fig. 2 for $n = 0.4, 0.5, 1, 2, \infty$ (exponential cooling). We also considered a model (not shown) which potentially allowed the cooling rate to be infinite initially, going over quickly to protracted cooling: the data chose parameters for this model such that it was practically indistinguishable from the $n = 1$ power-law shown. Despite the variety of cooling functions that could be represented, the data favors protracted cooling functions that are remarkably alike.

(1) Fast (but not infinitely fast) cooling from a high initial temperature $T_s \approx 4$ MeV, going over to protracted cooling after 2 seconds. This gives 14 and 5 for the most probable number of neutrinos to be detected in the Kamiokande and IMB detectors.

(2) Any cooling model is 10^5 times more likely than the constant temperature model.

(3) Exponential cooling ($n = \infty$) does not give the optimal best fit but agrees with the parameters obtained for exponential cooling in ref. 2.

(4) Models with protracted cooling ($n < 1$) give, however, better fit than the exponential model. For these models, the total energy $E_0 = 10^{53}$ erg is insensitive to n, but $N_0 = 5 \times 10^{56}$ $(n - 1/3)^{-1}$ $\bar{\nu}_e$.

(5) The best fit is obtained for $T(t) = T_s (1 + at/n)^{-n}$ with $n = 0.4$ 1/2 ⁖
$T_s = (4.2) \pm 0.5$ Mev $a = \tau^{-1}(0) = 0.14 \pm 0.05$ s. $N_0 = (1.3 \pm 0.4) \times 10^{58}$ $\bar{\nu}_e$
$E_0 = (9.1 \pm 3.3) \times 10^{52}$ ergs. Because of the protracted cooling, the energy radiated is 20% more than with exponential cooling and 50% more than with constant temperature.

(6) The time-averaged temperature $<T> = T_s (3n-1)/(4n-1) = \frac{1}{3} T_s$
$= 1.4$ MeV for $n = 0.4$ and $\frac{3}{4} T_s = 2.9$ Mev for exponential cooling. This makes the time averaged energy of emitted neutrinos $E_0/N_0 = 3.15<T> = 4.4$ Mev for $n = 0.4$ and 9.1 Mev for exponential cooling. Because Kamiokande and IMB detect only neutrinos above ~6.5 Mev and ~20 Mev, respectively, the time averaged energies of the neutrinos they observe are much higher, 12 Mev and 32 Mev respectively.

The continuous cooling models enable us to fit the Kamiokande clock to IMB time. From Fig. 3, we see that the Kamiokande data stream starts 0.1 ± 0.5 seconds after the first IMB event. As remarked, the last two IMB events then serve to fill in the otherwise unlikely 7 second gap at Kamiokande.

In summary, our continuous cooling model fit to all the Kamiokande + IMB neutrino events gives

(1) a good description of the entire data set, much better than any constant temperature model;

(2) a sequence of arrival times and neutrino energies consistent with blackbody radiation from a semi-degenerate Fermi gas of time-averaged radius $R = 35 \pm 10$ km.

(3) a number fluence decreasing initially at the rate $3a = 0.42$ seconds. The core cooling time is initially $(3a)^{-1} = 2.4$ seconds, in reasonable agreement with the time $\frac{1}{2}\frac{R^2}{\lambda c} = (\frac{M}{M_\odot})^{2/3} \rho_{14}$ for the diffusion of heat out of a neutronized core of mass $M \sim 1.4\, M_\odot$ at nuclear density $\rho = 2.4 \times 10^{14}$ g cm^{-3}, by $\nu_e + n \to \nu_e + n$ scattering with mean free path $\lambda = 51\, \rho_{14}^{-5/3}$ cm at core neutrino energies $\varepsilon = \frac{5}{6}\mu_e$ for neutrinos in β-equilibrium with degenerate electrons of Fermi energy $\mu_e = 184\rho_{14}^{1/3}$ Mev.

IV. Can Additional Time Structure Be Seen in the Data from SN 1987a?

We have just seen that the 19 Kamiokande + IMB neutrinos originate from a thermal source cooling by neutrino diffusion from an initial temperature $T_s = 4$ Mev rapidly down to 3.3 Mev in the first 2 seconds; thereafter as an $n \sim 1/2$ power law, slower than exponentially. Is there any additional time structure in the data that is statistically significant? Theoretically[3,4] a period $T \sim 0.1$ second might be expected if the neutrinos are being radiated from a pulsating neutrinosphere at density $\rho \sim 10^8$ g cm^{-3}. Other authors[3] claim to see a 8.91 ms period characteristic of a fast pulsar. Using two different statistical methods we have been unable to discern any significant periodicity in either the Kamiokande or IMB data separately. Because of the uncertainty in relative timing, we could not combine the IMB and Kamiokande data samples.

A broad band (0-100 Hz) analysis of the Kamiokande power spectrum (Fig. 4) shows power peaks no more significant than those generated by 11 points chosen at random over a 12.439 second interval (Fig. 5). Similar narrow band (100-125 Hz) analysis and separate analyses of the IMB data found peaks no more significant than those in random data sets of 11 events over 12.439 seconds and of 8 events over 5.6 seconds.

Searching for periods with about 0.1 seconds and in the 5-10 ms interval minimum residuals, we found residuals no more significant than those generated by random data sets. Indeed, we found in the Kamiokande data (Table 3) more than 20 periods with lesser residuals than the residual found for the 8.91 ms period reported[9].

V. Conclusions from Supernova 1987a.

We have found (Table 1) that the best joint fit to all the Kamiokande + IMB data is by a source cooling from an initial temperature T_s = 4.1 $\pm$0.5 Mev emitting E_o = (9 $\pm$ 3) x 10^{52} ergs and N_o = (1.3 x 0.4) x 10^{57} $\bar{\nu}_e$ on a neutrino diffusion time scale. This protracted cooling produces 50% more energy than the much less likely constant temperature model in Sec. II.

Assuming all the neutrino energy is equipartitioned among 3 flavors of ν_i and $\bar{\nu}_i$, the total energy release E_{TOT} = (5 $\pm$ 2) x 10^{53} ergs. This energy release is consistent with the binding energy (3 $\pm$ 1) x 10^{53} ergs expected for neutron stars. Because E_{TOT} is on the high side, this suggests collapse of a large iron core ~1.4 $M_\odot$ with a soft nuclear equation of state. Collapse of so massive a core is consistent with the delayed (revived shock) mechanism for supernova explosions, but not with the prompt (bounce shock) mechanism.

The number of ν_i + $\bar{\nu}_i$ flavors must be <8 and is probably ~3. No additional neutrinos, WIMPS, axions, Majorons, neutrino magnetic moments are required by the data. No significant periods are observed in any time scale.

The Kamiokande + IMB observations thus confirm the general features of the neutrino transport mechanism for supernovae: gravitation collapse of a degenerate iron core to a neutron star with binding energy released by neutrino transport on a weak interaction diffusive time scale. The data is too scant to test the details of any particular theory. Indeed, realistic calculations have still not produced a strong explosion by either the prompt or the delayed mechanism for neutrino momentum transfer to matter.

VI. Prospects for a Galactic Stellar Collapse.

A. Goals

The Kamiokande + IMB data revealed the diffusive cooling of a hot neutron star on a diffusive time scale but was too sparse to reveal any finer details of the pulse shape that would be diagnostic for the neutrino explosion dynamics. The same analysis in Sec. IV that revealed no significant time structure in the present 19 events would, in fact, become most significant with a fluence of 1000 neutrinos.

We have carried out the same power spectrum analysis described in Sec. IV for a simulated Galactic supernova at 10 kpc yielding ~500 neutrinos in simulated Kamiokande + IMB data, assuming that this time absolute timing is obtained at both detectors. We modulated our best fit cooling curve by a factor $(1 + 0.1 \cos 2\pi t/T)$ with period $T = 0.1$ seconds, persisting for 15 seconds (Fig. 6), for 5 seconds (Fig. 7) and for 1 second (Fig. 8). Figures 6 and 7 show that such a 10% modulation persisting through 150 cycles or through 50 cycles is perceptible at 10 Hz in the power spectrum. If this modulation persists for only 10 cycles, Fig. 8 shows that its 10 Hz power peak disappears into the noise. From this we conclude that features of the time signature lasting more than one second will probably be resolvable in a galactic supernova distant $D = 10$ kpc from Earth.

The goals of a Galactic supernova watch should include for **supernova physics**

(1) Resolving the neutrino explosion mechanisms: prompt, delayed or what-have-you.

(2) Studying the Kelvin-Helmholtz contraction and fall-back of accreting matter that is expected to produce neutron star heating before diffusive cooling.

For **particle physics** we would like to

(3) Study ν masses, lifetimes and mixing angles inaccessible in the laboratory.

Provided they are stable, neutrinos that are light enough to be produced in supernova can only have masses subject to the cosmological bound $m_\nu < 30$ eV. For neutrinos of energy E_ν (Mev), a finite mass m_ν(ev) leads to an arrival time delay

$$\Delta t = 0.25 \left[\frac{m_\nu (ev)}{E_\nu (Mev)}\right]^2 \left(\frac{D}{5 \text{ kpc}}\right) \text{seconds}.$$

This time delay will be measurable if enough flavor-i neutrinos $\overset{(-)}{\nu}_i$ of mass (1-10)eV are produced at supernova energies $E_\nu \sim 10$ Mev at Galactic distances $D \sim 5$ kpc. Some improvement may be expected in measurements of the ν_e mass, but if the masses of $\overset{(-)}{\nu}_{\mu,\tau}$ are in the cosmologically interesting region < 30 eV, they can be measured only over a long enough astronomical flight path. These $\overset{(-)}{\nu}_{\mu,\tau}$ are distinguishable from $\bar{\nu}_e$ statistically by their isotropic scattering off electrons. Since their detectable flux is about (1/70) that of $\bar{\nu}_e$, they can be studied once an intense enough source is produced by a nearby supernova.

B. Frequency of Collapse of Massive Stars

The Supernova 1987a neutrinos were recognized only after the supernova was discovered optically. The frequency of type II supernova in our Galaxy is only 0.01-0.1 per year, about the same as the pulsar birth rate. Most of these supernova are optically obscured by dust in the Galactic plane. The optical display is, in any case, a literally superficial phenomenon depending on conditions outside the collapsing core.

The rate of dying of massive stars (>10 $M_\odot$) in our galaxy is uncertain, the highest figure quoted being 0.09 star deaths per year in our galaxy[10]. With the experience and confidence gained from SN 1987a by neutrino physicists, we can expect astrophysical neutrino bursts to be recognized in the future whether or not accompanied by an optical display and whether or not a pulsar is left behind.

References

1. S. A. Bludman and P. J. Schinder, Ap. J. March, 1988.

2. D. N. Spergel, T. Piran, A. Loeb, J. Goodman, and J. N. Bahcall, Science 237, 1471

3. R. Mayle, Ph.D. Thesis, University of California, Berkeley, 1986.

4. J. Wilson in Numerical Astrophysics (J. M. Centrella, J. M. Le Blanc and R. L. Bowers, eds., Boston: Jones and Bartlett Publishers, 1985).

5. J. N. Bahcall, T. Piran, W. H. Press and D. N. Spergel, Nature 327, 682 (1987)

6. A. Burrows and J. M. Lattimer, Ap. J. 318, L63 (1987).

7. D. N. Schramm Comments Nucl. Part. Phys. 17, 239 (1987).

8. W. T. Eadie, D. Dryard, F. E. James, M. Ross and B. Sadoulet, Statistical Methods in Physics (Amsterdam: North Holland, 1971).

9. M. Harwit, P. L. Biermann, H. Meyer and I. Wasserman, 1987 preprint.

10. J. N. Bahcall and T. Piran, Ap. J. 267, L77 (1983).

Table 1

$$T = T_s\left(1 + \frac{at}{b}\right)^{-n}$$

n	a (s^{-1})	T_s (MeV)	N_0 (10^{58})	E_0 (10^{52} erg)	max(J) (10^3)
0.335	0.176	4.20	38.06	11.87	6.304
0.35	0.167	4.19	4.14	10.94	6.376
0.4	0.144	4.16	1.31	9.16	6.380
0.5	0.116	4.08	0.78	8.03	5.940
1.0	0.077	3.95	0.53	7.05	4.075
2.0	0.062	3.88	0.49	6.86	3.066
∞	0.052	3.84	0.46	6.69	2.197

Table 1. Maximum joint likelihood fit of the parameter $a = \tau^{-1}(0)$, T_s, N_0 of the cooling power law models described in Sec. III. The last column gives the (unnormalized) maximum likelihoods for each model. All of these models predict 14 + 5 for the most probable number of neutrinos to be detected in a Kamiokande and an IMB detector respectively. Actually observed were 11 + 8 neutrinos.

Table 2

Rank	P (ms)	R(P)
1	7.5434	0.427
2	7.5435	0.428
3	9.7568	0.430
4	9.7569	0.431
5	9.7570	0.432
6	9.7571	0.434
7	9.7572	0.438
8	9.7573	0.441
9	9.7574	0.446
10	9.7575	0.452
11	9.7576	0.458
12	9.7577	0.466
13	9.7578	0.473
14	9.7579	0.482
15	9.7580	0.491
16	9.7653	0.497
17	9.7654	0.497
18	9.7655	0.498
19	9.7656	0.500
20	9.7657	0.502

Table 2 The minimum value of R(P) for all periods P between 5-10 ms sampled in 0.1 μs steps, for the 11 Kamiokande events. Comparing with Figure 7, we see that these values are just what is expected for a random distribution. The 8.91 ms period reported by Harwit, Biermann, Meyer, and Wasserman is even less significant (*i.e.* has an even higher value of R(P)).

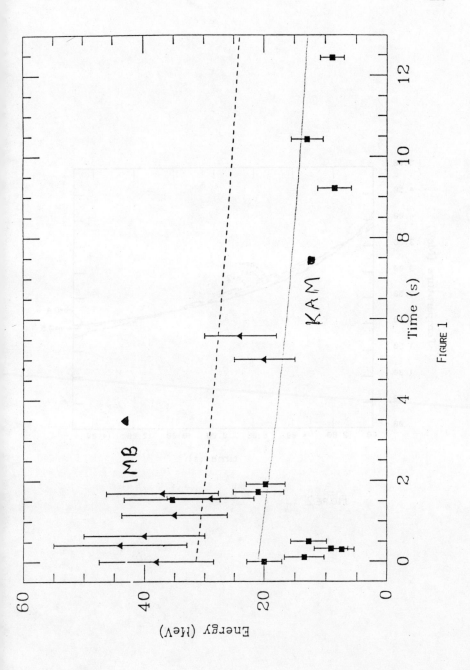

FIGURE 1

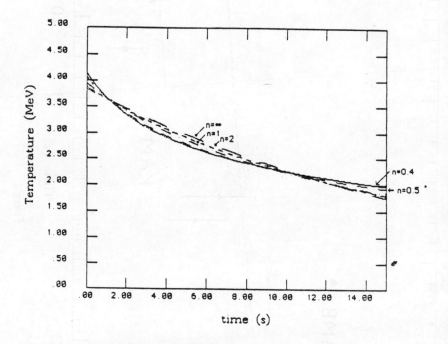

FIGURE 2

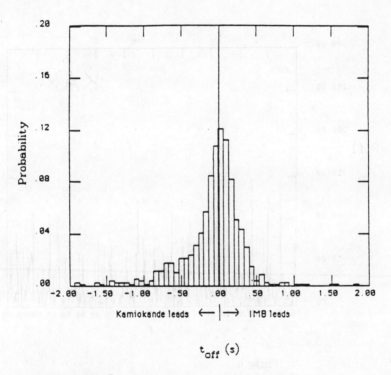

FIGURE 3

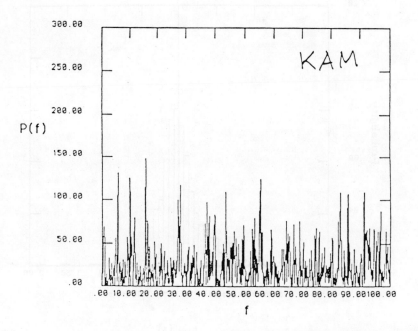

FIGURE 4

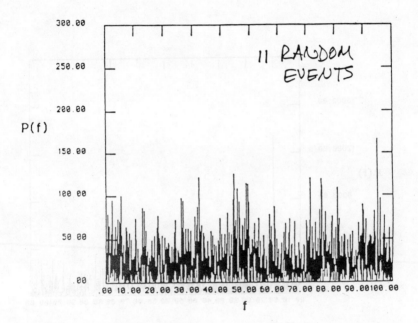

FIGURE 5

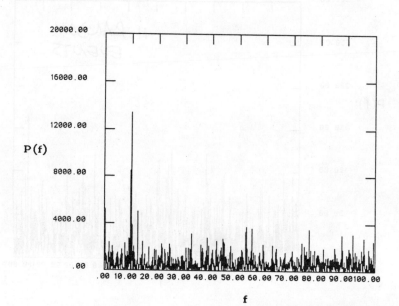

Figure 6

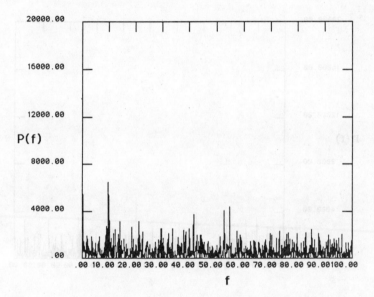

Figure 7

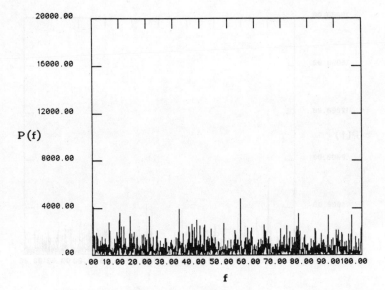

Figure 8

ULTRA-HIGH ENERGY NEUTRINO INTERACTIONS AND COMPOSITENESS

G. Domokos and S. Kovesi-Domokos
Department of Physics and Astronomy
The Johns Hopkins University
Baltimore, MD. 21218

Abstract. We review the arguments leading to the suggestion that muon-rich extensive air showers (EAS) and underground muon bursts associated with point sources in the sky may be caused by neutrinos rather than photons. If quarks and leptons possess a substructure with a characteristic energy scale Λ, neutrinos are capable of interacting with cross sections much larger than predicted by the standard model, once the CMS energy exceeds Λ.

We give estimates of the energy and angular distributions of hard "anomalous" neutrino interactions. Experiments are suggested in order to distinguish between photon and neutrino primaries.

1. Why Neutrinos?

All what follows is based on the assumption that quarks and leptons possess a substucture. We claim that we may have seen the first experimental hint at the existence of such a substructure, If someone can prove (???) that such a substructure does not exist, this paper becomes irrelevant.

It is by now well known that EAS associated with certain point sources in the sky appear to contain many more muons than expected. Moreover, muon bursts have been observed in underground detectors, associated with point sources. In a sense, it would be an insult to the participants of this Workshop, if we (theorists!) tried to review the situation. The explanation is still obscure, see, e.g.[1] and references quoted there, although it is our impression that the experimental evidence for the existence of some anomaly is getting more robust[2]. The standard argument runs as follows. The particles reaching us from a point source are obviously neutral, they are light, (lighter than, say, a GeV, although more stringent limits have been claimed) because we see a time structure.They must have interactions with matter comparable with strong interactions, otherwise, we could not explain the frequency of the anomalous showers or underground bursts, respectively. (We are aware of the fact that the reliability of the mass limits has been questioned[3] recently. Thus, clearly, more work is needed to clarify the question. However, our intuitive feeling is that even at present, the totality of the data suggests a light primary.) All this together means trouble. We cannot have a light hadron (or so we are told by our colleagues manning accelerator-based experiments). Photons cannot cause the effect, unless by some new physical mechanism, one can "turn off" pair creation[4]. Neutrinos are out of the question <u>within the framework of the standard model</u>, as a back-of-an-envelope estimate quickly demonstrates, and supersymmetric models are not faring much better either[5].

However, one of us together with S. Nussinov pointed out that if there exists a substructure going beyond presently recognized quarks and leptons, then we may be able to endow neutrinos of a sufficiently high energy with an anomalously strong interaction[6]. The qualitative argument runs as follows. Every "reasonable" composite model of quarks and leptons must assign some common constituents to quarks and leptons: otherwise, the model has no right to exist and we may just as well stay with the quarks and leptons themselves as fundamental constituents of matter. Let us now imagine that we managed to construct a good composite models. Then, presumably, leptons (neutrinos in particular) contain strongly interacting constituents. As soon as we reach an energy regime where the available CMS energy exceeds some characteristic energy, Λ, of the model, one can produce particles of $P_T > \Lambda$, say. Thus, the substructure is resolved and large - P_T processes will be dominated by the strong interactions of the subconstituents. In particular, we gave a rough estimate of the inclusive cross section of the reaction,

$$\nu + N \to \mu + X,$$

where N stands for any nucleon; we used a crude parton model to connect the cross section on nucleons to that on quarks. We found in this way that the large cross sections claimed are possible, although, one indeed needs some new physics in order to understand the data.

This explanation has, in our opinion, the advantage that one does not have to invoke a mechanism for "suddenly" (as compared to "logarithmically") "turning off" a known interaction (of which one has seen no example, but it is conceivable). In fact, as we have mentioned it before, the "new physics" has no competition in ν-induced reactions. Furthermore, as we shall try to argue, the proposed scenario is further testable even at its present, embryonic, stage.

A potential disadvantage is that further theoretical investigation may show that a consistent theory with the properties we believe must be postulated, just cannot be constructed. This is, however, hardly a reason to abandon the hypothesis before the investigation is finished.

2. Review of the Neutrino Inclusive Results

We are interested in the reaction in which an incoming ν dissociates into two groups of preons; one of the groups carries the leptonic quantum numbers, the other one is strongly interacting. The dominant process at high P_T (where the preon structure can be resolved) will be one where the strongly interacting group of preons exchanges gluons with a target quark. (We are deliberately vague about details of the process, a "group of preons" may, in fact, be one single preon, etc.) The crucial point, however, is that we assume that <u>two</u> energy scales are relevant in the energy range under consideration. The lower one is $\Lambda_{QCD} \simeq 0.1$ GeV, the other one is the preon scale, which is expected to be $\Lambda \simeq 1$TeV. The important observation made in ref.6 and then verified in ref.7, was that under such circumstances:

i) The inclusive cross section is proportional to the <u>total</u> cross section of the strongly interacting preon group on the target quark (call it σ_0) and, hence it is of $O(\Lambda_{QCD}^{-2})$: it is "big".

ii) Once the energy in the CMS is large enough in order to have $P_T \geq \Lambda$ sufficiently often, the coefficient of σ_0 scales: it depends on dimensionless variables such as the energy fraction, z, of the observed muon and its transverse momentum fraction, x_T.

Let us recall that in terms of quantities observed in the LAB frame, to a good approximation,

$$z = E_\mu/E_\nu ,$$
$$x_T = 2 P_T (s)^{-1/2},$$

where, as usual, s stands for the CMS energy squared. In refs.6 and 7 we gave approximate analytic expressions for the inclusive cross section, $d^2\sigma/dz\, dx_T$ in terms of a crude, but acceptable, approximation to the nucleon structure function; in fact, we just took $F(x) = (1-x)^3 /x$, which is qualitatively right.

We now have made some minor improvements on the representation of the structure functions, using information available in the 1986 edition of the standard "Particle Properties Data Booklet". No serious attempt has been made to derive more accurate expressions of the neutrino-quark inclusive cross section: given the present status of composite models, any "improvement" could only convey a false impression that we know our physics better than we actually do.

Without further ado, we thus present our results. Figs.1(a) thru 1(d) sample the x_T distibution for various values of z. In Fig. 2, we give the distribution in x_T, integrated over z. On reading the Figures, one should remember that no "high-P_T filter" (see refs. 6,7) has been folded in: thus, depending on the value of $4\Lambda^2/s$, one should, effectively, cut off the distribution for $x_T^2 \leq 4\Lambda^2/s$. For the range of the scaling variables which is of interest, it is convenient to plot the quantity,

$$(\sigma_0)^{-1} x_T\, d^2\sigma/dz\, dx_T,$$

since our knowledge of σ_0 is quite uncertain: the renormalization group calculation, first introduced in ref.6, gives only a crude estimate. (In particular, with the leading order estimate we carried out, one has no idea about the size of confinement effects. One can only argue on an intuitive basis that such effects should be small.) The fact that we have factored out x_T, reflects the fundamental assumption that the group of preons interacting with the target quark does not carry any characteristic scale other than the ones mentioned before. The important assumption in all these calculations is that the projectile dissociation amplitude does not decrease rapidly with the four-momentum squared of the preon group exchanged between the ν and the target quark.

That quantity can become quite large; from kinematics, it is given by the expression,

$$t \approx - s\, (x_T^2/4z).$$

Thus, it is significant that the data can be understood only if one assumes no rapid decrease of the virtual dissociation amplitude with t.

It seems that one may have to revise the concept of "compositeness", see ref.8. (We thank Larry McLerran for discussions on this subject.)

3. Further Tests: Cosmic Absorbers, etc.

If indeed neutrinos are responsible for the μ-rich EAS and the μ-bursts in underground detectors, one should have an entirely new regime of neutrino astronomy starting at $(s)^{1/2} \approx 1\text{TeV}$; some of this has been outlined in ref.9. In particular, if high energy neutrinos have an anomalously large (from the standpoint of the standard model) cross section in matter, one should observe an otherwise inexplicable deficiency of high energy neutrino-induced reactions coming from underfoot in underground detectors as compared with the neutrino spectrum when the source is overhead. It seems that at one time there was such a claim: at the 20th Rencontre de Moriond, Bionta presented some data acquired by the IMB collaboration, which indicated that CYG X3 was "seen" by IMB when it was overhead, but not when it was underfoot. We have been told[11] that the effect disappeared when more data were analyzed. However, given all the uncertainties, we feel that another look at the data may be justified. In particular, it would be interesting to look at HER X1 during the burst (≈June '86) claimed by the LANL-U.of MD. group, see ref.2.

One can devise a different kind of test of these ideas, based on the following considerations.

A number of point sources has been identified in the sky, all of them "visible" (at least in principle) to UHE detectors. The sources so identified are at distances from us varying from a few kpc to about 50 kpc. It has been pointed out by Gould and his collaborators[12] that UHE photons should be absorbed rather efficiently by the cosmic photon gas. (In particular, Gould, in his letter to Ap.J., quoted in ref.12, called attention to the fact that the data taken by the Kiel group on "photons" coming from CYG X3, should not smoothly extrapolate the spectrum derived from X-ray data. It is interesting to speculate that Gould's remarks provide the first, prima facie evidence for a neutrino-induced muon excess.)

Keeping this in mind, we propose that one should look (by means of both underground detectors and EAS arrays) at point sources located at varying distances. In this way, one puts varying amounts of thickness of the most important "cosmic absorber", i.e. the "cosmic photon gas", background, between the source and the terrestrial detector. The proposed test is, in principle, an extremely simple one.

α) It is known, cf. Gould & al., ref.12 that photons are absorbed rather efficiently by the cosmic photon background. By contrast, we found no process in the framewwork of any reasonable preon model leading to a comparable absorption of a ν beam.

β) Consequently, if a "distant" source is "seen" in EAS arrays by means of μ-rich showers or in underground μ-bursts, the effect is (almost) certainly caused by neutrinos. By contrast, if the effect seen is due to a new hadron or some unexpected behavior of photons, nearby sources, such as HER X1, or CYG X3 should be "visible", but distant ones should not.

One can find a simple semiempirical formula for the mean free path (mpf) of a photon in the cosmic photon gas, by interpolating the results of Gould and Schreder, ref.12. In the energy range of (approximately) 2 PeV $\leq$ E $\leq$100 PeV, we find:

$$\lambda/10 \text{ kpc} \simeq 0.7 \text{ (E/PeV)}^{0.25}.$$

Below we list the approximate distances of some relevant binary sources.

TABLE I.
DISTANCES OF SOME BINARY SOURCES

SOURCE	DISTANCE/10kpc
LMC X-4	5
CYG X-3	1.3
HER X-1	0.4

By looking at the Table, we see that, Hercules is within one mfp for the entire relevant energy range; Cygnus is marginal and LMC definitely should NOT be seen in underground detectors or in muon-rich showers if those are caused by photons. (Assuming LMC X-4 to be comparable to, say, CYG X-3 at high energies - which may be a very optimistic assumption - one expects an event rate reduction by a factor of $\simeq$15, which makes the experiment difficult but potentially feasible. On the same basis, we may expect the ν flux from any of these sources to be comparable to the γ flux.

Acknowledgement

This research was supported in part by the U.S. Department of Energy, under Grant No. DE-FG02-85ER40211. We have started this investigation during our stay at the Aspen Center for Physics. During our stay there, we greatly benefitted from tutorial sessions on pulsar physics given to us by S. Colgate. We also benefitted from discussions with L. McLerran and S. Bludman.

Footnotes and References

1. See, e.g. F.Halzen, in Proc. SLAC Summer Inst.on Particle Physics,Ed. E.Brennan. SLAC, Stanford, CA. 1987
2. G.Yodh, invited talk, Johns Hopkins Workshop on Heavenly Accelerators, March 1987.
3. J.R. Cudell, F.Halzen and P.Hoyer, Phys.Rev. **D36**, 1657 (1987).
4. This has, indeed, been proposed, cf. F. Halzen, P.Hoyer and N. Yamdagni, Phys. Lett. **190B**, 211 (1987).
5. J.R. Cudell and F. Halzen, Phys.Rev. **D36**, 346 (1987) and references quoted there.
6. G. Domokos and S. Nussinov, Phys.Lett. **187B**, 372 (1987).
7. G. Domokos and S.Kovesi-Domokos, preprint JHU-HET 8610 (1986).
8. G. Domokos, preprint JHU-TIPAC 8710 (1987).
9. G. Domokos and S.Kovesi-Domokos, preprint JHU-HET 8706 (1987).
10. R.M. Bionta, in Proc. XX. Rencontre de Moriond, Les Arcs, March 1985.
11. R. Svoboda, private communication.
12. R.J. Gould and G.P. Schreder, Phys. Rev. **155**, 1404 and 1408 (1967).
    R.J. Gould, Ap. J. **274**, L23 (1987).
13. H. Lee and S.A. Bludman, Ap.J. **290**, 28 (1985).
    T.K. Gaisser and Todor Stanev, Phys. Rev. Letters, **54**, 2265 (1987).

Figure Captions

Fig 1, (a) through (d). Plot of the inclusive cross section for the reaction described in the text, for various values of the energy transfer to the muon.

Fig.2. Plot of the transverse momentum distribution for the same reaction.

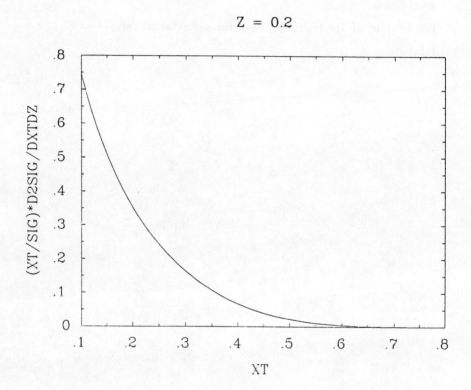

Figure 1(a)

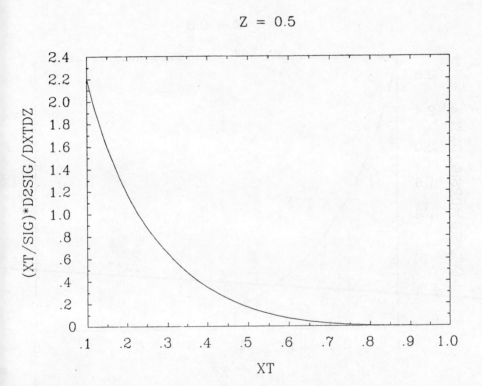

Figure 1(b)

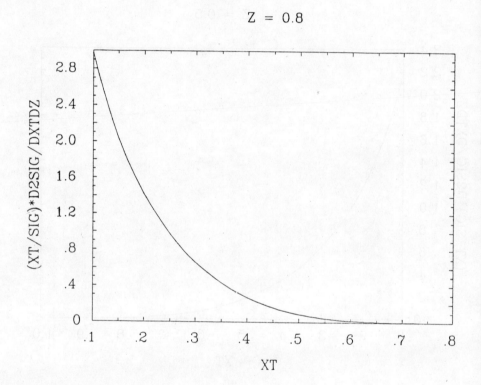

Figure 1(c)

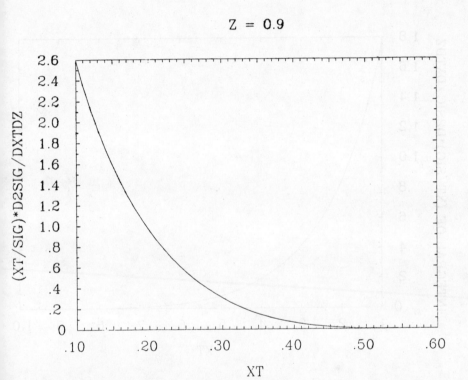

Figure 1(d)

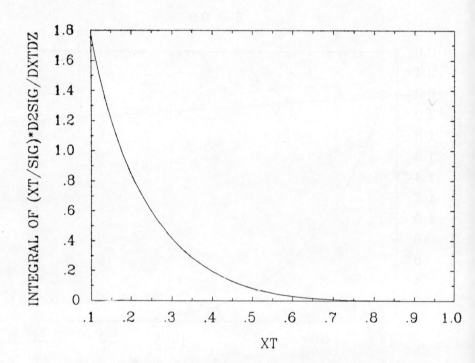

Figure 2

NEW DETECTORS FOR
SUPERNOVA NEUTRINO BURST OBSERVATION

David B. Cline

Physics Department
University of California, Los Angeles
405 Hilgard Avenue
Los Angeles, California 90024-1547

ABSTRACT

Two types of extremely massive detectors capable of recording extra galactic SN neutrino signals have now been proposed: (1) very large water detectors (possibly in a high lake – HLD or LENA), and (2) large detectors to observe the neutron emission from neutral current interactions. We briefly review these two kinds of detectors, commenting on the relative merits of each.

1. Introduction

The detection of the time difference between neutrino bursts from distant supernova is likely the only way to measure neutrino masses in the eV to few eV range. Since neutrinos of this mass provide significant cosmological effects, these observations would have profound consequence. Likewise, the existence of neutrino mass would have great significance for elementary particle physics. The various types of neutrinos from the SN event $(\nu_e, \bar{\nu}_e, \nu_\mu/\nu_\tau \cdots)$ carry different information concerning the dynamics of the collapse and thus such observations may lead to a better understanding of the mechanism of the stellar collapse as well.

For these reasons a study has been going on at UCLA for the past year to evaluate different types of massive neutrino burst detectors. This study has already yielded one new idea (to be described in Section 3) for a suitable detector.

2. Very Large Water Detectors (HLD or LENA)

These detectors use the reactions

$$\bar{\nu}_e + p \rightarrow e^+ + n$$

or

$$\nu_x + e \rightarrow \nu_x + e$$

and detect the Cherenkov radiation from the relativistic electrons. A new generation of water detector is being studied for the detection of such events.

The concept of ocean or lake detectors was popularized by the DUMAND group during the 1970's (the idea was apparently due to C. Cowen originally). During the first two DUMAND workshops there was considerable discussion of the possiblity of detecting supernova neutrino bursts. However, the dynamic range required to detect 10 MeV and 100 GeV neutrinos was too large and DUMAND was further designated as a high energy neutrino and muon detector. Recently, several new studies of lake detectors have been carried out for the detection of muons from neutrino interactions outside the detector (LENA, GRANDE, HLD). We have applied these concepts to the detection of supernova bursts. One detector that is being modeled is a 250,000 ton lake detector that uses about 10,000 of the larger phototubes used in the Kamiokande detector (called the HLD, high lake detector, Figure 1). With such an array of PMT's, the threshold for the detector is between 15 − 20

MeV. The Super Kamioka detector will use a similar number of PMT's for a mass of 30,000 tons with an expected threshold of (7 − 10) MeV.

These two studies illustrate the trade off between the lowest threshold and the detector mass. The HLD detector is designed to collect the largest number of ν_μ and ν_τ interations from a galactic SN in order to determine the ν_μ and/or ν_τ mass. Super Kamiokande on the other hand, is optimized for $\bar{\nu}_e$ interactions which are expected to have a lower energy spectrum of the ν_μ and ν_τ. Recently M. Koshiba has considered a detector with a variable density of PMT's toward the detector center. A sketch of this detector is shown in Figure 2.

The major uncertainty in the construction of any SN detector is the expected rate of SN in the galaxy. A 2nd problem is the actual likelihood of extracting a neutrino mass from the resultant neutrino signal due to the long time spell of the neutrinos ($\sim$ 10 − 20 sec) and the relatively small expected time difference for $\nu_e, \bar{\nu}_e$ neutrinos.

(for example, if $E_\nu = 10$ MeV, $M_{\nu_e} = 10$ eV

distance 10 Kpc, $\Delta\tau \simeq 4$ seconds)

Since ν_μ and ν_τ are likely to have a larger average energy, a similar situation exists even for $M_\nu \sim 30$ eV, for example

($E_{\nu_\mu} = 20$ MeV, $M_{\nu_\mu} \simeq 20$ eV, 10 Kpc and $\Delta\tau = 4$ sec)

Thus the galaxy is almost too small to be used as a time of flight spectrometer to measure (10 − 20) eV neutrino mass and, coupled with the possible long wait for another supernova (10 − 50 years), leads to somewhat discouraging prospects.

3. A New Method for the Detection of Distant Supernova Neutrino Bursts

Reproduced in appendix A is a recent paper proposing a new type of detector that uses the neutral current reaction

$$\nu_x + \text{Nucleus (N, Z)} \rightarrow \text{Nucleus}^* (N-1, 2) + n + \nu_x$$

with the detection of the neutron using an inexpensive detector. If such detectors could be made up to the mass of $10^7 - 10^8$ tons, they would likely revolutionize the search for neutrino bursts from SN in other galaxies. Several steps are now required to investigate the use of these detectors

(a) More complete calculations on the neutral current reactions and backgrounds.

(b) A study of radioactive backgrounds – various types of materials (such as $CaCO_3$).

(c) Measurements of the neutral current cross sections.

(d) Study of inexpensive, massive neutron detectors.

If these steps (a – d) prove successful, it will be suitable to propose one or more such massive detectors.

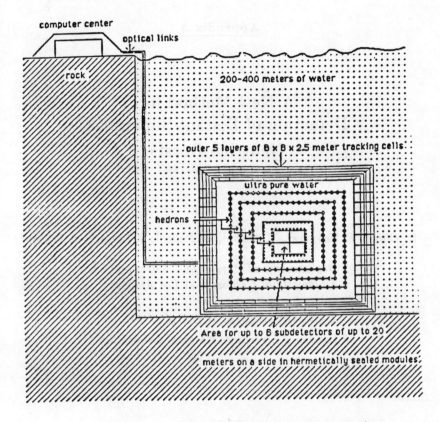

Figure 1. The HLD Detector Concept (D. Joutras)

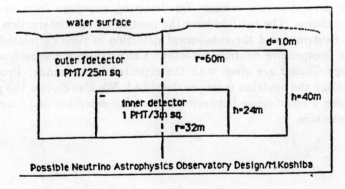

Figure 2. The LENA Concept (M. Koshiba)

Appendix A

A New Method for Detection of Distant Supernova Neutrino Bursts

D. Cline[1], E. Fenyves[2], T. Foshe[1], G. Fuller[3], B. Meyer[3], J. Wilson[3]

1 Departments of Physics and Astronomy
University of California at Los Angeles

2 Department of Physics
University of Texas at Dallas

3 Lawrence Livermore National Laboratory

ABSTRACT

We present a new concept for detecting neutrinos from very distant stellar collapse. The method uses the neutral current interaction of the ν_μ and ν_τ neutrinos and the subsequent detection of time correlated neutrons with an inexpensive neutron detector. Calculations for supernovae out to the Virgo Cluster are given with the expected backgrounds. Prospects for determining the neutrino mass are described. We also discuss the possibility of forming a coincidence between the neutrino detection and future gravity wave detectors.

The recent detection of the neutrino burst from SN 1987A by the Kamiokande and IMB groups provided an important verification of the theory of stellar collapses and a new limit on the electron neutrino mass.[1,2,3] Unfortunately the expected rate of galactic supernova events is from 10^{-1} to 100^{-1} years. Thus future supernova neutrino detectors require the observation of extra galactic stellar collapses. It is expected that the Virgo Cluster will have several stellar collapses per year. The development of a method to detect the neutrinos from these events on Earth is the goal of the work reported in this paper. The distance to the Virgo Cluster is estimated to be 20 Megaparsecs, and the different time of flight of massive ν_μ or ν_τ neutrinos would allow precise mass measurements or determination of upper limits on the masses if the mass of either neutrino is greater than about 1 eV.

The commonly accepted modes of detecting neutrinos used either the inverse β decay reaction

$$\bar{\nu}_e + p \rightarrow e^+ + n \qquad (1)$$

or the elastic scattering reactions of neutrinos and antineutrinos

$$\nu_e + e \rightarrow \nu_e + e \qquad (2)$$
$$\nu_{\mu,\tau} + e \rightarrow \nu_{\mu,\tau} + e \qquad (3)$$

These reactions either require free protons in the detector (reaction 1) or have very small cross sections (reactions 2, 3). Reaction 3 is the only reaction of the three that is available to μ and τ neutrinos. We propose to use instead of (3) the following inelastic scattering of neutrinos through neutral current interation

$$\nu_x + \text{Nucleus}(A, Z) \rightarrow \text{Nucleus}^*(A - 1, Z) + n + \nu_x \qquad (4)$$

A key element in our scheme is the expected relatively high energy spectrum of the ν_μ and ν_τ neutrinos in the stellar collapse. Note that the cross section per nucleon of the target for reaction (4) is about one order of magnitude greater than that for reaction (3) for neutrinos of the same energy well above the threshold energy for reaction (4). This leads to a large reduction in the mass of the detector required to see μ and τ neutrinos from the Virgo Cluster of galaxies. ν_e neutrinos generated in supernovae have lower energy spectrum below the threshold energy of reaction (4).

Another key element in the proposed scheme is the detection of the neutrinos from reaction (4) using inexpensive – massive neutron detectors in particular a mineral deposit.

We now describe recent calculations of some of the details of the technique discussed here. The cross section for neutral current interaction on ^{40}Ca is shown in Figure 1a. Note the rapid rise of the cross section above 15 MeV neutrino energy ($T_{\nu_\mu,\nu_\tau} = 5$ MeV). The cross sections per nucleon for carbon and oxygen are close to that of calcium. The estimated average cross section for neutron production off ^{40}Ca for neutrinos from a typical stellar collapse is (the mean μ, τ neutrino energy in collapse is 30 MeV and the corresponding $T_{\nu_\mu,\nu_\tau} = 10$ MeV) 4.0×10^{-42} cm^2 or 1.0×10^{-43}cm^2 per nucleon. This is about one order of magnitude larger than the cross section of reactions (2), and (3) per nucleon of the target, and approximately equal to the cross section of reaction (1) per nucleon of the target for 10 MeV $\bar{\nu}_e$'s in a water detector. Thus the detection of a supernova using the neutral current reaction could have the same overall efficiency as that of conventional water detectors if the neutrons from reaction (4) can be detected with the same efficiency as the positrons from reaction (1).

We compute the cross section for excitation of ^{40}Ca to an energy of 24.15 MeV (corresponding to two oscillator levels) by inelastic neutrino-nucleus scattering via neutral current interaction with a model based on a scaled-up version of charged-current interactions.[4] This cross section multiplied with a neutrino energy distribution emitted from a black body at T = 10 MeV is shown in curve $\sigma \times f_\nu$ of Figure 1b. The excited ^{40}Ca nucleus is then allowed to decay via proton, neutron, or gamma emission. We calculate that the probability to emit a neutron is $\Gamma_n = 0.17$. The spectrum of neutrons emitted is shown in Figure 1c.

Our estimate of the cross section for inelastic neutrino scattering is based on a simple fit to Haxton's (1988) shell model calculations. In principle it may be possible to use electron scattering experiments to get electromagnetic form factors for the various nucleii. A minimum of theoretical input then allows a calculation of the appropriate weak neutral current form factors (Donnelly 1973, Walecka 1975, Fukugita 1988, Haxton 1988). Thus, eventually, it should be possible to get accurate, semi-experimental results for the inelastic cross sections of all the target nuclei.

We consider for illustration a possible detector configuration using $CaCO_3$ as the neutrino interaction medium (Figure 2). Cylindrical BF_3 detectors of raduis r_1 are inserted into the $CaCO_3$ material and the neutrinos are detected by the reaction

$$n + B^{10} \to Li^7 + He^4 \qquad (5)$$

A neutron thermalizing material (CH_2) is inserted into the $CaCO_3$ material around the BF_3 detector with radius r_2. We have calculated the efficiency of neutron detection by varying the parameters r_1, r_2 and the radius r_3 of the $CaCO_3$ that is the primary neutrino target. A detailed neutron propagation computer program for all the materials was developed at LLNL and used for the calculation. For example, the calculations give an overall neutron detection efficiency of 20% for the array of Figure 2, with $r_1 = 20$cm, $r_2 = 22$cm, and $r_3 = 60$cm. We can now calculate the number of detected neutrons ($N_{detected}$ or N_{counts}) from a supernova at distance R (Megaparsecs) for an overall detector of volume V (m^3) of $CaCO_3$ including the efficiency for neutron detection assuming $N_\nu = 6 \times 10^{57}$ from the supernova and $\sigma_n \sim 2.0 \times 10^{-43}$cm^2 per nucleon for $CaCO_3$ by the following expression

$$N_{detected} = \frac{2 \times 10^{-5}}{R^2} V$$

For supernovae in our galaxy (R $\simeq$ 0.02 Megaparsecs) we find

$$N_{counts} \simeq 0.05 \times V$$

A modest detector of size 10m × 10m × 10m detects about 50 counts. These counts occur in a time interval of (10 – 20) seconds, and this signal should be easily detected above the background. For the local cluster of galaxies

$$R \simeq 1$$

and we find

$$N_{counts} \simeq 2 \times 10^{-5} V$$

A detector of 100m × 100m × 100m gives about 20 counts, again a signal that is likely to be above background. In order to observe supernovae in the Virgo Cluster we require a detector of about 10^8 tons (e.g., 100m × 1000m × 1000m). Improvements in neutron detector efficiency could reduce the required mass somewhat. We have also considered NaCℓ and SiO$_2$ as possible detector media. NaCℓ is somewhat poorer regarding neutron transport than CaCO$_3$, but is better than CaCO$_3$ for neutron production by the inelastic neutrino scattering. SiO$_2$ is much better than CaCO$_3$ as regards neutron transport, and about the same for neutron production.

We now turn to the question of backgrounds. The most important characteristics of the neutron detection is its time correlation to the neutrino burst. We have calculated the time of propagation of the neutrons in the detector and find that 90% of the neutrons are counted in the first millisecond after production. The neutrino signals are expected to have durations of about ten seconds with the intensity peaked towards early times. Thus, the detector time response is quite adequate. However, because the neutrino signal is spread out over several seconds it is necesary to have a very small neutron background rate.

The key background rejection technique would use the slow coincidence counting of neutrons from the neutrino interactions in more than one BF$_3$ detector when combined with the absence of counts within a millisecond in other parts of the detector (to reject neutrons produced by nuclear photo excitations initiated by high energy cosmic ray muons) and the correct time structure of the event. In order to estimate the rate of background events from the heavy element impurity in the CaCO$_3$ we use recent measurements of the neutron flux at the Gran Sasso Laboratroy in Italy (the Gran Sasso Mountain is largely made of CaCO$_3$)[5]. From these measurements we estimate that the neutron count rate in a 10 m length of BF$_3$ detector is $(10^{-2} - 10^{-3})$ sec^{-1}. For a 10^5 ton detector the average rate is $(1 - 10^{-1})$ sec^{-1} and the signal from a galactic supernova is far above background.

The net time separation of the neutrino bursts from ν_1 and ν_2 (where ν_1 is assumed to have a negligible mass) for neutrino energy E$_\nu$ and from a stellar collapse at distance D can be written

$$\Delta \tau = 41 \text{ sec } \left(\frac{D}{\text{Kiloparsec}}\right) \left(\frac{M_\nu}{100 \text{eV}}\right)^2 \left(\frac{10 \text{MeV}}{E_\nu}\right)^2$$

For a neutrino mass of cosmological significance (M$_\nu$ = 10 eV) and < E$_\nu$ > = 30 MeV we find

$$\Delta\tau = 46 \text{ sec} \quad \text{for } D = 1 \text{ Megaparsec}$$

Thus the detection of a stellar collapse at about 1 Megaparsec would be sensitive to a neutrino mass of (10 – 20) eV. At a lower mass the star will have to be correspondingly farther away. Such a collapse could be detected with a megaton detector.

There is going to be considerable improvement in gravitational wave detectors in the near future for both the large bar detectors and the laser interferometer detectors.[6] These detectors may in the next decades become sensitive to stellar collapse in the local cluster, perhaps even out to the Virgo Cluster. In this case we suggest to form a coincidence between gravitational wave detectors and the detectors of a correlated neutrino neutron burst using the technique proposed here.

Most stellar collapse events are thought to be optically invisible because of dust obscuration. In addition, stars in the 8 to 10 $M_\odot$ range may eject much of their envelope by stellar winds and may be optically weak. The standard estimated rate of stellar collapses in our galaxy is 0.01 per year, which is inferred from the historical supernovae. Rates as high as 0.1 per year have been estimated from studies of stellar populations.[7] Neutron star collapse will also be a good source of neutrinos and is probably a good source of gravitational radiation as well. In order to understand the events that are unobserved optically, their detection by both gravity waves and neutrinos is clearly desirable. If neutrino masses are several tens of electron volts, then the neutrinos from distant galaxies will be spread out in time such that recognition of their signal will be difficult. If, however, a gravity wave is detected, then using that event to determine the initial time will greatly enhance the chances of picking up the neutrino signal. Conversely, if a clear neutrino signal is observed, the search for a gravity wave in the detector signal would be more reliable.

We have illustrated the technique in this letter. Detailed calculations and study of various types of natural materials must be carried out before the construction of such a detector could be proposed. However, we believe our preliminary calculations demonstrate the feasibility of this approach.

ACKNOWLEDGEMENTS

We would like to acknowledge most valuable discussions with Drs. J. Bahcall, S. Woosley, J. Ferguson, R. Bauer, W. Haxton.

REFERENCES

1. K. Hirata, et al. Phys. Rev. Lett. <u>58</u>, 1490 (1987).
2. R. Bionta, et al. Phys. Rev. Lett. <u>58</u>, 1494 (1987).
3. For a review of the neutrino emission from Stellar Collapse and SN 1987A, R Mayle and J. R. Wilson, Ap. J. submitted 1987, and A. Burrows and J. Lattimer, Ap. J. (Letters) <u>318</u>, L63 (1987), and D. Joutras and D. Cline, Astro. Lett and Communications, 1988 Vol. 26, pp 341-347 (1988).
4. T.W. Donnelly, Phys. Lett., <u>43B</u>, 93 (1973), J. D. Walecka, in "Muon Physics", vol 2, ed V. W. Hughes and C. S. Wu (Academic Press, New York, 1975), M. Fukugita, Y. Kohyama, K. Kubodera, IAS, preprint, 1988, W. Haxton, U. W. preprint, 1988.
5. E. Bellolti, et al. "New Measurements of the Rock Contamination and Neutron Activity in the Gran Sasso Tunnel", preprint, Univ. Milano, (1985).
6. E. Amaldi, private communication.
7. J. N. Bahcall and T. Piran, Ap. J., <u>267</u>, L77, (1983).

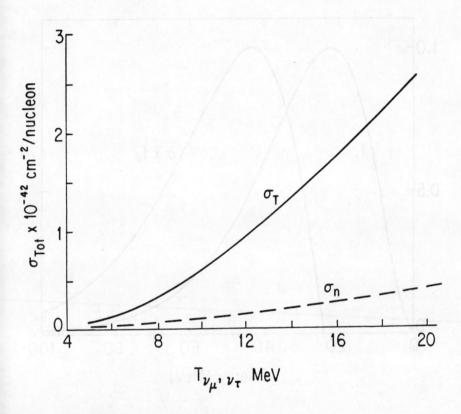

Figure 1a. $\nu_\mu, \nu_\tau + {}^{40}$Ca cross sections per nucleon for total neutral current scattering (σ_T) and single neutron production (σ_n), averaged over neutrino black body spectra at temperature T_{ν_μ,ν_τ}.

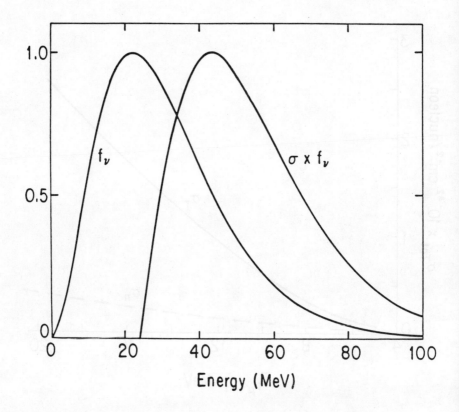

Figure 1b. Neutrino spectrum, f_ν, from a black body at T = 10 MeV and total inelastic neutrino scattering cross section off ^{40}Ca multiplied by f_ν versus neutrino energy adjusted to peak at one.

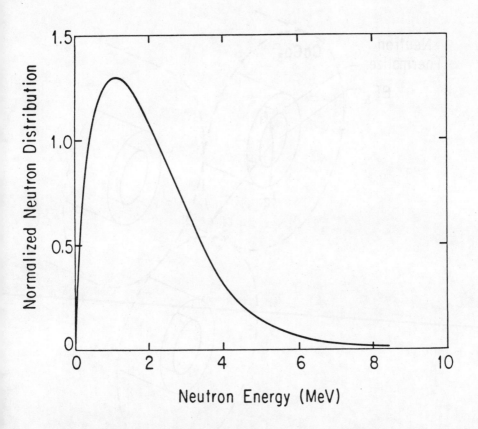

Figure 1c. Normalized energy spectrum of neutrons emitted from excited ^{40}Ca.

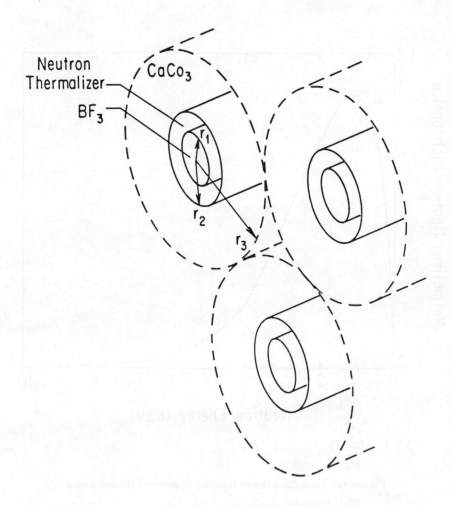

Figure 2. Distant supernova neutrino burst detector.

A "New" Old Neutrino-Detector Concept

Peter C. Bosetti

III. Physikalisches Institut, RWTH Aachen, West Germany

Abstract

A simple method is proposed for the construction of a large Supernova Neutrino Telescope using water cherenkov technique. Several modules of this type could be sensitive to all supernovae within our local group of galaxies. Other physics and astrophysics capabilities of the detector are briefly discussed.

Introduction

It seems appropriate to divide the next generation of neutrino-astronomy into three categories (neglecting the very low-energetic neutrinos from the big bang), first the high energies (>100 GeV), originating from point sources. Detectors for such neutrinos should have an effective area of the order of 100.000 m². Secondly there are the medium energy neutrinos (1 - 10 GeV) from the atmosphere, which could be used to detect the matter oscillations. Here an area of at least 10.000 m² seems appropriate. The last category are the low energy neutrinos (5 - 50 MeV), coming from the sun and supernovae. Clearly, detectors for such neutrinos should be as large as possible. In this note I want to discuss the possibility to build a Supernova Neutrino Telescope.

Aim of the detector

The aim of a dedicated Supernova Neutrino Detector should be to have a reasonable chance to observe at least 2 supernova bursts within the lifetime of the detector, i.e. roughly 10 years. Unfortunately, the rate of supernovae is rather small and predicted to one every 20 to 50 years for a galaxy like our own. Clearly, a supernova neutrino detector should be able to monitor galaxies beyond our own, at least all galaxies within our local group. There are some 20 objects in the local group, most of them smaller than our galaxy. M31 at 2.2 lightyears is the closest one comparable to ours. This implies a detector size at least 170 times as big as those that have observed SN 1987A, i.e. at least 170.000 m³.

The essential quantities to be measured are the number and arrivaltimes of the neutrinos and, preferably, the direction of their origin. As the positron from the inverse beta-decay, the most prominent reaction of the antineutrinos, does not reflect the direction of the neutrino, the last requirement can only be fulfilled, if the detector consists of at least three modules, separated by at least 20 km. with a timing accuracy of 10^{-6} s this would result in a determination of the direction of the supernova to about a degree, if it occured within our galaxy.

The "New" Detector

Building "conventional" water cherenkov detectors for neutrinos from supernovae like IMB and Kamiokande, a cube of 55m^3 with a surface of around 18.000 m^2 to be equipped with photomultiplier, would be necessary. Going to several modules will result in a worse surface to volume ratio, even though such a detector system would be desirable. In order to reduce the construction costs of such a modular system, we propose to build several cylinders with only top and bottom plates equipped with photomultiplier. The cylinder wall should consist of reflecting material. Diameter and height of the cylinder can be optimized to have a minimum number of light reflections before the photons reach either top or bottom, leaving the radius as small as possible, so that comparably few photomultipliers are needed for the light detection. 6 to 8 such modules should be constructed resulting in a volume of 180.000 to 210.000 m^3.

A detailed Monte Carlo study is necessary to find the optimum parameters for the modules. First results show that a radius of 15m and a height of 40m appear promising. The average number of scatters is rather low, and the area to be equipped with photomultipliers, assuming 6 of such 30.000 m^3 modules, is 8400m^2, a substantial saving against conventional modules.

In fig. 1, some typical events in one of the modules are shown. The neutrino energy is assumed 10 MeV, 20 MeV, and 10 Mev in a, b, and c, respectively. Furthermore, 90% of the light produced is assumed to be reflected by the wall. Note that only every 10th photon is really generated. It can be seen that the lightcone produced is not completely destroyed by the scattering at the wall, or can at least be reconstructed. Fig. 1c shows a very unfavourable event for such a detector, namely one that is produced in the middle with the positron going towards the wall. However, even in this case the event is not lost, and the direction can be reconstructed (with large errors).

As the main purpose of the detector is the detection of supernova neutrinos, background events are much less severe than searching for solar neutrinos or proton decay, the signature being several (>10) events within a few seconds. The detction efficiency for 10 MeV neutrinos is already high (>75 %) and approaches 100% at 20 MeV.

It should be noted that the detector would have seen around 1500 to 2000 neutrinos from SN 1987A, and would see more than 10 neutrinos from a similar supernova in M31.

Physics and Astrophysics Capabilities

The main purpose to build such a detector system is, of course, the observation of neutrinos from supernova-bursts. Such observations could provide new limits on neutrino masses, their lifetimes, but also important insight into the physics of collapsing stars. However, a detector of the proposed type is as well capable to search for other phenomena. We just mention a few here:

– proton decay

whether or not the detector would be sensitive to nucleon decay depends crucially on the location, i.e. the background from muons and neutrino interactions in the corresponding energy-range.

– upward going muons

the detector can have an area of around 10.000 m^2, opening the possibility to search for point sources of neutrinos.

– neutrino mass limits

in the energy-range between 0.5 and 1 GeV the flux of atmospheric neutrinos is expected to be different due to neutrino matter oscillations, whether or not the neutrinos have to pass through the earth's core.

– axions and axion-like particles

limits on fluxes of pseudoscalar particles predicted in electroweak theories would come mainly from an accurate measurement of the energy output from a supernova going into neutrinos.

– monopoles catalyzing baryon decay

the signature for monopoles catalyzing baryon decay is unique. Several times during a few microseconds an energy depostion of about 1 GeV will occur, while the monopole passes through the detector.

Conclusions

Without any doubt a Supernova Neutrino Telescope, i.e. a device seeing more than one supernova during its lifetime, is desirable. The design of such a detector is clearly governed by the available funds. A "conventional" water cherenkov detector with its surface completely equipped with photomultiplier would be useful, however such a detector(-system) will be very expensive and might thus never be built.

The 6 – 8 cylindrical modules proposed here will reduce the costs substantially without loosing to much of the quality of the detector. The proposed detector not only would be sensitive to neutrinos from a supernova anywhere within our local group of galaxies, it would add in addition some other interesting physics possibilities. To search for differences in the flux of 0.5 to 1 GeV/c neutrinos of atmospheric neutrinos going upwards and downwards (i.e. the search for matter – oscillations using the earth's core) is one example.

The "ultimate" aim for a Supernova Neutrino Telescope is being sensitive up to 36 lightyears, i.e. the Virgo cluster. From the 2500 galaxies in this cluster one can expect one event approximately every 3 days.

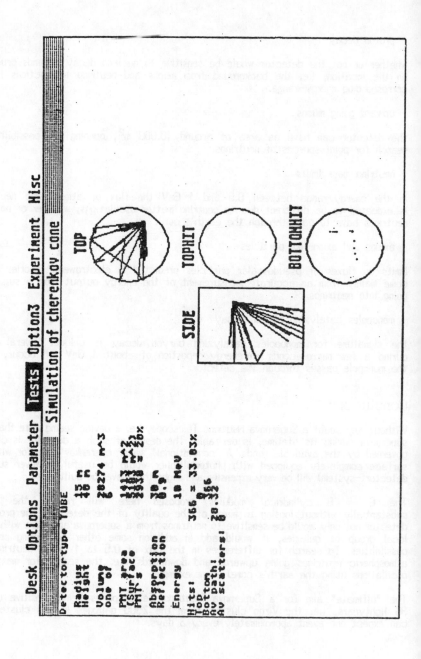

Fig. 1 a. Simulation of a 10 MeV neutrino

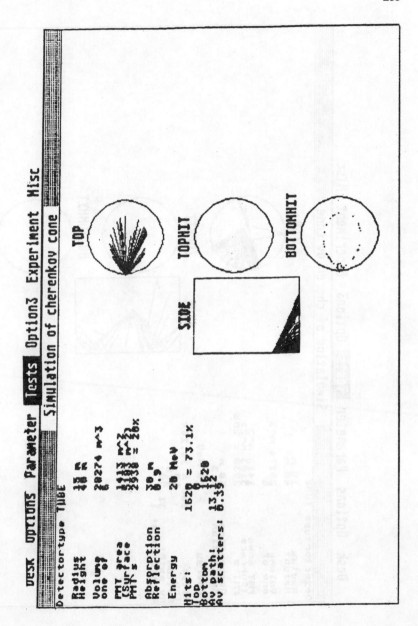

Fig. 1 b. Simulation of a 20 MeV neutrino

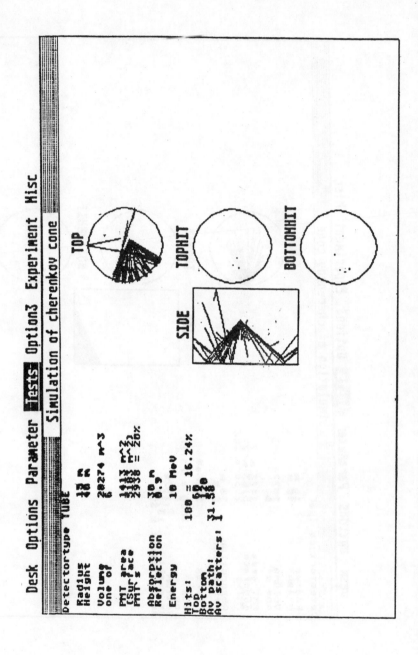

Fig. 1 c. Simulation of a 10 MeV Neutrino

GRANDE

A GAMMA-RAY AND NEUTRINO DETECTOR

R. Ellsworth,[1] W. Gajewski,[2] J. Goodman,[3] J. Gaidos,[4]
T. Haines,[2] W. Kropp,[2] J. Learned,[5] R. Novick,[6] M. Potter,[2]
F. Reines,[2] J. Schultz,[2] H. Sobel,[2] R. Svoboda,[2]
A. Szentgyorgyi,[6] C. Wilson,[4] and G. Yodh.[3]

1) George Mason University
2) University of California, Irvine
3) University of Maryland, College Park
4) Purdue University
5) University of Hawaii, Manoa
6) Columbia University, New York

Abstract

A proposal to build a detector facility primarily for neutrino and gamma-ray astronomy is described. Unique features of this detector include operation at the surface of the earth and a very large surface area (6 x 10^8 cm^2) which is 100 percent active for the electromagnetic, hadronic, and muonic components of extensive air showers.

The technology for operating large, ring-imaging, water Cerenkov detectors has been perfected over the last several years with the successful construction and operation of the underground IMB proton-decay detector[1]. This detector has been able to search for rare events (a few per year) having characteristic topologies, in a background of 10^8 per year. Based on this experience, we believe that it is now reasonable to build an even larger water Cerenkov detector at the surface of the earth. Such a detector would be able to search for sources of extra-terrestrial neutrinos. Additionally, the unique approach to air shower detection afforded by this device would allow one to study the suspected sources of high energy gamma-rays in great detail, and to resolve the controversy concerning the muon component of the showers associated with these gamma-rays.

Paper presented by H. Sobel

A collaboration has been formed to build and operate such a particle detector facility, combining a neutrino telescope and an extensive air shower array (EAS). The device would have a large area, 6.2×10^8 cm^2. The energy threshold (angular resolution) would be <10 GeV (~ 1°) for neutrinos and ~ 10 TeV (<0.5°) for gamma-rays. The detector would be located at the surface of the earth in a water-filled pit.

The proposed design, construction, and operation of this new detector is based largely on the technology developed for IMB and on commercially-available reservoir technology. Essentially no new developmental work is required.

All previous large detectors for neutrino astronomy have been built deep underground to shield them from cosmic-ray backgrounds. Using the results from IMB and other detectors, we have been able to demonstrate the feasibility of operating this detector at the earth's surface[2].

The conceptual design of the facility is illustrated in Figure 1. It consists of two essentially independent detectors. While each could be built and operated separately, combining the two in the same facility enhances the physics potential, while permitting a considerable savings in time and money. Additionally, the interaction of the two detectors is constructive, e.g., the EAS can act as an effective anti-coincidence for the neutrino telescope.

The Neutrino Telescope:

The field of neutrino astrophysics was begun in an exciting fashion with the observation of a burst of low energy neutrinos from SN1987A[3]. It has been speculated that supernova remnants, X-ray binaries, and many other objects may be sources of both high-energy neutrinos and gamma rays[4]. Our detector will be two orders of magnitude larger than the largest of the currently operating detectors (IMB at 400 m^2), and therefore will represent a significant increase in sensitivity in the search for high-energy extraterrestrial neutrinos.

The neutrino telescope (Fig. 1) consists of three optically-isolated layers of downward-facing photomultiplier tubes (PMT's). High-energy neutrino interactions in the rock below the detector produce upward-going muons which penetrate the detector. If we include the weak-interaction kinematics as well as muon scattering in the rock, 90% of the muons are within 2.6° (5°) of the parent neutrino direction for a neutrino flux with a spectral index of 3.0 (2.3). Thus, a detector angular resolution between 1° and 2° is appropriate.

The design of a large, surface detector must allow for its operation in the presence of the vast numbers of cosmic-ray muons. For example, a 250 x 250 m^2 plane 30 m below the water surface is penetrated by about 1.4×10^6 muons/sec. Data from

IMB shows how such operation is possible. Fig. 2, scaled from IMB data, gives the relative illumination of downward-facing PMT's by penetrating muons in one layer of the proposed detector. A simple cut on the number of illuminated PMT's yields a factor of about 100 in up-down discrimination.

A detailed monte carlo simulation of the proposed three-layer detector employing a realistic cosmic-ray shower parameterization shows that the trigger rate for the full size detector can be limited to a few per second[2].

The real background for this, as well as any other earth-bound neutrino telescope, is the flux of neutrinos produced by high-energy cosmic-ray primaries interacting in the atmosphere. We have carried out a detailed calculation[2] of the atmospheric-neutrino background expected in the detector, and find that the total rate would be about 16 per day. Figure 3 shows the expected atmospheric-neutrino background as a function of declination for several representative detector sites. Also shown is a selection of potential sources of high-energy neutrinos. If we take as an example Vela X-1 observed from a Southern California site, atmospheric neutrinos contribute 840 events/yr/sr; a 3° cut about the source would then include about 7 events/yr. If we define the minimum detectable flux to be a 3 σ enhancement in five years, this flux would correspond to a minimum sensitivity of 8 x 10^{-16} events/cm^2/s. The flux from Cygnus X-3, a well-studied source, is estimated to be several times this value[7].

A total of about 6000 atmospheric-neutrino events per year would be detected; this is about six times as many as has been seen, to date, by all other detectors. With this large signal, many questions of current interest events can be studied.[5]

The Gamma-Ray Telescope

The gamma-ray telescope (Fig. 1) consists of two optically isolated layers of upward-facing PMT's. The upper layer, about 10 m deep, would serve as a total absorption calorimeter for the showers. All charged particles above Cerenkov threshold radiate light which would be recorded by the PMT's; additionally, photons in the shower, not normally observed by most existing arrays, would convert in the 10 meter layer so that their energy would also be measured.

The basis of this detector is quite different from that of any other existing or planed air shower array. Traditional arrays sample only a few percent of their total area. The observations are then converted, statistically, into an equivalent number of particles for the shower. Our detector has continuous coverage over its entire area; the Cerenkov light can be converted into energy deposition in the detector instead of number of particles. These differences should reduce problems associated with finding the hadronic core(s) and with fluctuations in sampling statistics. Additionally, having continuous coverage would facilitate studies of the lateral distributions of the various shower components.

The lower layer of the gamma-ray telescope would be about 20 to 30 m below the surface, deep enough so that only muons would be observed. This part of the detector, also 100% sensitive, would allow detailed studies of the shower's muonic component.

According to the conventional model, gamma-ray initiated showers should have only about 5 to 10 percent of the number of muons observed in proton-initiated showers of the same total energy. It is currently believed that this difference can be exploited to discriminate against proton showers, which are the major background for gamma-ray astronomy. However, recent results from the Cygnus Air Shower Array at Los Alamos[6], as well as earlier works, e.g. that of the Keil group[7], suggest that that signals from both Cygnus X-3 and Hercules X-1 have a larger-than-expected muon content. This puzzling result underscores the critical need to understand the muonic component of the showers, and shows the value of a new generation of detectors, designed to study the detailed properties of the electromagnetic, hadronic, and muonic components of the shower.

REFERENCES

1. R. Bionta, et al., Phys. Rev. Lett. **51**, 27 (1983).

2. W. Gajewski, et al., Proposal for a Design Study of a Surface Cerenkov Detector to Investigate Astrophysical Sources and Particle Interactions, UCI 87-4, January 1987.

3. R. Bionta, et al., Phys. Rev. Lett. **58**, 1494 (1987).

4. T.K. Gaisser and T. Stanev, Phys. Rev. Lett. **54**, 2265 (1985); V.J. Stenger, AP. J. **284**, 810 (1984); G. Cocconi, Phys. Lett. (1986).

5. R. Svoboda, et al., Ap. J. **315**, 420 (1987); J. LoSecco, et al., Phys. Rev. **D35**, 2073 (1987); J. LoSecco, et al., Phys. Rev. Lett. **54**, 2299 (1985); Various Papers, IMB Collaboration, Proc. 20th International Cosmic Ray Conference, Moscow (1987).

6. G. Yodh, et al., Highlight Talk, Proc. 20th International Cosmic Ray Conference, Moscow (1987).

7. M.A. Samorski and W. Stamm, Ap. J. (Lett.) **286**, L17 (1983).

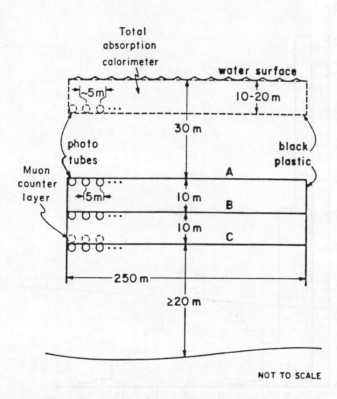

Figure 1: Diagram of a possible configuration of the gamma-ray (dashed lines) and the neutrino telescope (solid lines).

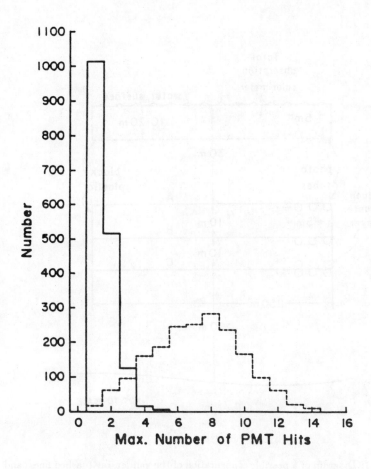

Figure 2: Maximum number of PMT hits in a 20 ns coincidence window for top (solid line) and bottom (dashed line) planes of the IMB detector (scaled for a 5 meter tube lattice).

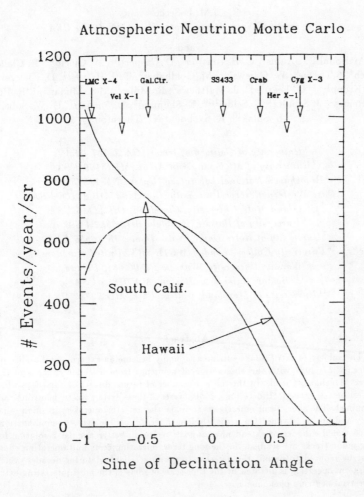

Figure 3: The expected background due to upward-going muons from atmospheric neutrinos as a function of source declination for two hypothetical detector sites.

Future Directions for the IMB Detector

J.M. LoSecco
University of Notre Dame, Notre Dame, IN 46556 USA
and
Z.J. Ajaltouni[7], J. Annis[6], C. B. Bratton[5], D. Casper[2,10], A. Ciocio[10], R. Claus[10],
M. Crouch[4], S.T. Dye[6], W. Gajewski[1], M. Goldhaber[3], T. J. Haines[11], D. Kielczewska[9,1],
W. R. Kropp[1], J. G. Learned[6], J. Matthews[2], R. Miller[1], M.S. Mudan[8], L.R. Price[1],
F. Reines[1], J. Schultz[1], S. Seidel[2,10], E. Shumard[2], D. Sinclair[2], H. W. Sobel[1],
L. Sulak[10], R. Svoboda[1], G. Thornton[2]

[1] *University of California, Irvine, CA 92717 USA*
[2] *University of Michigan, Ann Arbor, MI 48109 USA*
[3] *Brookhaven National Laboratory, Upton, NY 11973 USA*
[4] *Case Western Reserve University, Cleveland, OH 44106 USA*
[5] *Cleveland State University, Cleveland, OH 44115 USA*
[6] *University of Hawaii, Honolulu, HI 96822 USA*
[7] *University of Notre Dame, Notre Dame, IN 46556 USA*
[8] *University College, London WC1E 8BT, United Kingdom*
[9] *Warsaw University, Warsaw PL-00-681, Poland*
[10] *Boston University, Boston, MA 02215 USA*
[11] *University of Maryland, College Park, MD 20742 USA*

Abstract

The subject of extra solar neutrino astronomy became an experimental reality on February 23, 1987 with the observation of neutrinos from a stellar collapse in the Large Magellanic Cloud. In this note a number of future directions for the world's largest detector, the IMB detector a codiscoverer of these extra galactic neutrinos, are examined. The short term objective is to make the detector operate with high reliability and low dead time. On a slightly longer term, possible detector modifications which would enhance or extend physics goals requiring better angular resolution for long muon tracks are studied. In the long term, operating costs and operating effort must be reduced to permit the detector to produce reliable data for decades. Such long term operation is dictated by the time scale of some of the most interesting extra terrestrial neutrino phenomena.

1. BRIEF HISTORY.

The IMB detector grew out of concerns in 1979 to study the stability of matter. Theories clearly indicated that proton decay could occur with a lifetime of less than 10^{31} years. The imaging water Cherenkov detector was a novel idea that combined source and detector in a very low cost, easily scaled up to the 8000 ton total mass design. The only electronic subsystem of the detector was good time resolution Cherenkov light detectors.

High quality light transmission was a technological problem that through care in design and operation, was mastered. All reconstruction of events is based on the time of arrival, pulse height and pattern of hit photomultiplier tubes. A schematic view of the detector is shown in figure 1.

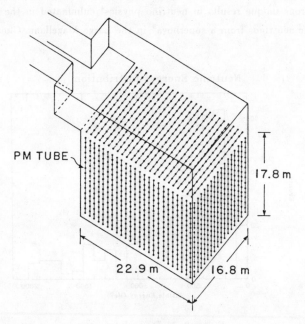

Figure 1: The IMB Detector

In about 1982 this detector produced definitive limits for proton decay into almost 30 different modes[1]. While the discovery of proton decay was not made the detector was an enormous success at setting significant limits far below original expectations. Early results used rather crude[1] but sensitive techniques, but as we became more sophisticated full scale event reconstruction[2] was introduced. While the detector resolution was not comparable to the best of high energy physics technology its enormous mass more than compensated for any inefficiencies. The problem of proton decay was quickly reduced to one of getting a good understanding of a very small (5%) background.

The initial operation of the detector lasted from August 20, 1982 to July 6, 1984. The first detector operated with 5 inch bare hemispherical photomultiplier tubes. A brief period from September 15, 1984 to June 25, 1985 saw operation with 5 inch tubes augmented by 2 foot by 2 foot waveshifting plates[3]. A significant modification, converting from 5 inch to

8 inch photomultiplier tubes with waveshifter plates, was completed by May 3, 1986 and the detector has operated continuously since then. This change increased light collection fourfold over the initial operation of the device.

While the physics program has always concentrated on proton and bound neutron decay, numerous unique results in neutrino physics[4] culminated in the observation of extra galactic neutrinos from a supernova[5] in the Large Magellanic Cloud in February 1987.

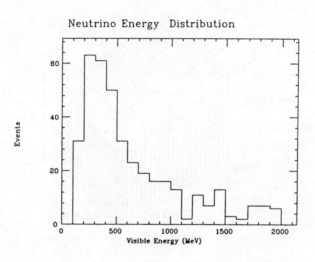

Figure 2: Neutrino Interactions

In figure 2 we illustrate the visible energy distribution of events observed in the IMB 1 detector. These events are believed to be of caused by atmospheric neutrinos and constitute a background to any astronomical search on or in the earth.

In the rest of this paper, we will explore modifications and improvements to the detector to permit it to continue to operate, as a competitive scientific instrument for years to come.

2. SHORT TERM FUTURE - LIVE TIME.

The major considerations for the next few years of operation of the IMB detector are high reliability and reduced dead time. Both of these objectives will give us more physics live time and hence more data.

The detector reliability has been high due to redundancy in most major systems. The detection of light is done with 2048 photomultiplier tubes with waveshifter plates (light collectors). They are uniformly distributed about the surface of the detector. The failure of any one of these has a negligible effect on the trigger threshold or the detection efficiency. We have always operated the detector at very low light levels so that the possibility of a Poisson fluctuation to zero at a photomultiplier has always been large. These fluctuations have the same effect as a failure of a small subset of the light collectors. So no modification of the software or hardware is needed to deal with routine failures. In general the detector operates with about 2% of the light collectors inoperative at any given time.

Because of this redundancy and also high reliability, pool maintanance that requires shutting down the experiment and entering the detector, can be restricted to one shift per month. The importance of the redundancy is emphasized with our supernova observations. 25% of the detector was turned off because of a high voltage power failure and yet the signal was unambiguiously observed and distortions introduced by the failure can be calculated.

The major sources of lost information at present are operator and computer related. Calibrations and detector studies still consume a significant fraction of our down time but the largest share is associated with computer tape writing and computer associated dead time. The computer associated dead time is currently about 16%. This can be improved. The ambient muon trigger rate and the digitizing time dictate a maximum required dead time of about 1%. Multihit electronics could reduce this to nominally zero but afterpulsing and the preponderance of hits on the bottom of the detector from throughgoing muons make this an uncertain option at best.

To deal with fluctuations in the ambient trigger rate of 2.7 Hz, we have employed a set of buffer memories. All information can be transfered to temporary storage and the detector made live within 1 msec. after the digitizing finishes. This means that any information processing can proceed at the average trigger rate but short bursts of information can be efficiently recorded.

Improvements to the system to reduce data acquisition dead time and tape writing could be combined. The cost of microprocessors is now very low. Events can be digitized directly into the memory of an ACP class computer[6] (68020 or equivalent) where all of our reconstruction algorithms can be immediately applied. Multiple processors can be employed to keep up with the workload. The final product would be a full fit to the event and the raw data could be discarded. This represents a factor of about 100 reduction in data output with no reduction in information extracted. Tape writing could be reduced to only one tape a month from the one every 17 hours, at best, now.

A number of specific schemes are currently under consideration. The Fermilab ACP project has produced a device suitable for our needs. It is commercially available from Omnibyte[6] in West Chicago. These devices are directly linked to VME which is a powerful new standard for computer interfacing. It is straightforward to connect the system to experiments either directly or through Camac[7]. Camac, at present, looks convenient but would be a bottleneck in some applications. In the IMB application we must transfer and analyze only about 25,000 bytes per second on the average. But our experience has emphasized the need to handle data bursts at rates up to two orders of magnitude, or more, higher than this. VME can keep up.

Simpler, less expensive, solutions to the immediate problem of 16% dead time are also possible and under active study. They will not fully process the data and have limited burst rates.

3. MIDDLE TERM FUTURE - TRACKING.

The current IMB detector has an angular resolution[8] of about 7° for tracks that traverse the detector. These tracks are typically due to downward muons that could be studied for point sources (muon astronomy) or upward going muons coming from muon neutrino interactions in the rock around the detector. These later are searched for point astronomical sources of neutrinos.

The 7° is an instrumental limit coming from the coarseness of the sampling. Intrinsic limitations would give a resolution of about 1°. Improving the detector angular resolution to this ultimate level would give an increase of signal to background for point source searches by nearly a factor of 50.

One way to improve the angular resolution for entering tracks is to install tracking chambers[9] around the detector volume. Inexpensive tracking chambers could easily yield this angular resolution by indicating the entry and exit points for such tracks in the detector. The cherenkov light could be used to resolve the sense ambiguity associated with purely tracking chamber measurements. Figure 3 illustrates how the IMB detector would look with tracking chambers surrounding it.

Among physics improvements would be better comparison with such detectors as Soudan II and Nusex for muon astronomy. IMB has approximately the same depth as the Soudan group but the poor angular resolution has made comparison difficult until now. Point source neutrino searches would of course also be improved. The more traditional IMB physics goals may also be improved. Detector calibrations could be easily cross checked with throughing muons. Muon associated dead time could be virtually eliminated since only the sense of the track need be determined from the Cherenkov detector and this could be easily done with patch scalers or QDC's. Only part of the event would need to be read in and processing would be very limited. Potentially the fiducial volume

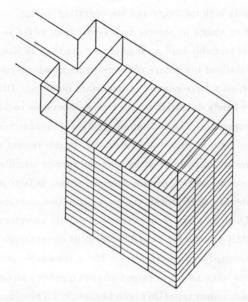

Figure 3: The IMB Detector with Tracking Chambers

could be extended. The current fiducial volume cut of 2 meters from the tube plane walls is dictated by both reconstruction errors and the attenuation of entering neutrons and photons. Tracking chambers could eliminate the reconstruction error from consideration since a track would be known to have come through a wall. Subsequent analysis of events near the walls will be needed to adequately determine the attenuation length for neutral backgrounds.

Even a subset of chambers at the top of the detector would achieve some of these goals. Half of our cosmic ray muons enter the detector through the top. Most of the problematic events occur near the corners of the detector. A partial shield would still permit calibration cross checks and would halve the work required to make a trigger decision. Caution is of course called for to avoid rejecting upward going muons that penetrate the top from below. This idea is under active consideration by many of the members of the IMB collaboration.

4. LONG TERM FUTURE - TRANSIENT PHENOMENA.

Any realistic search for galactic gravitational stellar collapse events must operate for periods in excess of 20 years to have a reasonable chance of seeing an event or even setting a realistic, interesting limit on the occurance of such phenomena in our Galaxy. This calls for large detectors with high reliability and low operating costs.

IMD has proved its ability to observe such events from as far as 50 Kpc. (A serious galactic search would probably need a range of no more than 20 Kpc.) The observations have in many ways clarified the nature of the neutrino in astrophysics. But in other respects, they have raised a large number of important questions. Discrepencies amongst experiments that are barely significant for the recent observation could be studied in great detail with a closer supernova. The angular distribution of the observed supernova induced events is unusual and by no means well understood. In this respect an imaging detector such as IMB can provide more information than the simpler scintillator tanks traditionally used for such searches. The traditional technique has, in fact, proved unreliable for SN1987A. Imaging detectors can provide the energy, direction and time of occurance of the individual events. These data can be used together to fully understand the phenomena.

To operate the IMB detector for 20 to 30 years in its current mode would be expensive and wasteful[10] of talented human resources. For a reasonable period, data from the detector will continue to be useful to extend physics questions already studied by IMB. The cumulative data summary tape (DST) can be consulted if new ideas emerge for physics with atmospheric neutrinos or other new phenomena. But soon the data sample will be so large that to extend it in any statistically meaningful way, such as a factor of 2, would require 5 to 8 years more running. This time scale pretty well precludes graduate student or post doctoral involvement.

A solution to this problem is to operate the detector with an effort commensurate with the incremental knowledge to be extracted. Each new year of operation now is worth only about 20% of our first year. Reduction of operating effort can be accomplished if a short term program is established to make the detector autonomous. Maintenance schedules can be reduced and operating as an unmanned device extended. The IMB detector presently operates unmanned for 80% of the time. This represents an even larger fraction of the useful livetime. Routine tasks can be eliminated or assigned to part time technicians.

Much of our physicist effort at present goes into data analysis. Only the last stages of this require detailed physicist intervention. Major reductions in operating cost would come about from eliminating computer tape handling entirely.

In the long run, the most significant observation to be made by IMB could well be of transient phenomena such as supernova neutrinos. The ambient flux of muons and neutrinos can be studied with existing data samples. Each additional neutrino observed increases our data sample by about .1%. It is worth a possible reduction in detection efficiency for these events to keep the detector in continuous operation for decades. Observation of a galactic supernova event would increase our knowledge and understanding manyfold over the current situation. Questions about stellar dynamics, neutrino mass and neutrino oscillations could be studied in detail rather than in speculation as at present. On the other hand the continuing effort devoted to this task can not be high since the likelihood of success in any given year is very low.

Because of this we are exploring the possibility of a long term, "dormant" operation of the IMB detector. Such a detector could continue to do programatic non-accelerator physics at a low priority while waiting for very rare transient phenomena to occur. We have learned from SN1987A that many detector details, such as efficiencies with only 75% of the detector operational, can be determined after the event of interest has been found and identified. The software developed for the current effort can be used for decades with no modifications.

One of course needs to ensure detector stability. The geology of the IMB site is very stable due, at least in part, to the innovative way the excavation was carried out. Stability of other components such as electronics and phototubes has been good. The redundancy mentioned above and the existence of adequate spare components would go a long way toward achieving this goal.

An in depth study is needed to adequately determine what changes would be required to guarantee reliable long term operation.

5. CONCLUSIONS.

All of the ideas discussed in this paper are provisional. The problems are of common concern but the specific solutions to them are under discussion. The costs of implementing these improvements are much less than the costs of new detectors that would pursue similar objectives. Beyond that the IMB team is knowledgeable and has been very successful in establishing the field of non-accelerator high energy physics and in exploiting it far beyond original expectations.

It is clear that the future of physics will depend on high quality instruments of the type discussed here and elsewhere at this conference.

6. ACKNOWLEDGEMENTS.

This work was supported in part by the U.S. Department of Energy. We are grateful to the Morton-Thiokol Company for the use of their Fairport Mine. One of us (JML) would like to thank the organizers for an invitation to participate in this workshop.

7. REFERENCES.

1. B. G. Cortez, et al., Phys. Rev. Lett. 52, 1092 (1984).
   H. S. Park, G. Blewitt, et al., Phys. Rev. Lett. 54, 22 (1985).
   G. Blewitt, J. LoSecco, et al., Phys. Rev. Lett. 55, 2114 (1985).
   T. Haines, et al., Phys. Rev. Lett. 57, 1986 (1986).

2. G. Blewitt, et al., Search for Two-Prong Proton Decays at IMB, Proceedings of the XXIII'rd International Conference on High Energy Physics, Berkeley, California (1986) pages 1290-1292, edited by S. Loken (World, Singapore, 1986).
   G. Blewitt, A Search for Free Proton Decay and Nucleon Decay in O^{16}, Using the Invariant Mass and Momentum of Exclusive Final States, Caltech Ph.D. Thesis October, 1985.
   S.C. Seidel, A Search for Nucleon Decay Employing Particle Identification, Invariant Masses, and Complete Track Reconstruction in a Water Cherenkov Detector University of Michigan Ph.D. Thesis April, 1987.

3. R. Claus, et al., A Waveshifter Collector for a Water Cherenkov Detector, April 27, 1987 (submitted to N.I.M.)

4. J. LoSecco, et al., Phys. Rev. Lett. 54, 2299 (1985).
   J. LoSecco, et al., Phys. Lett. B184, 305, (1987).
   J.M. LoSecco, et al., Phys. Lett. B188, 388, (1987).
   J. LoSecco, Phys. Rev. D35, 1716 (1987).
   J. LoSecco, et al., Phys. Rev. D35, 2073 (1987).
   R. Svoboda, et al., The Astrophysical Journal, 315, 420-424 (1987).

5. R.M. Bionta, et al., Phys. Rev. Lett. 58, 1494 (1987).

6. J. Hansen, Omnibyte, (West Chicago, Il.).
   Fermilab ACP project (Batavia, Il.).

7. V. Klotz, Bytech Engineering Ltd. (North Vancouver, B.C., Canada).

8. R. Svoboda, et al., The Astrophysical Journal, 315, 420-424 (1987).
   R.M. Bionta, et al., Phys. Rev. D36, 30 (1987).

9. J. Learned, Tracking Chambers Surrounding IMB, IMB internal report, September 19, 1987.
   J. LoSecco, Muon Tracking Chambers, IMB internal report, March 26, 1987.

10. J. LoSecco, The Long Term Future of the IMB Detector, IMB internal report, July 9, 1987.

Cosmions and Stars

P. Salati *

Miller Research Fellow at the University of California at Berkeley.
Theoretical Physics Group, Bldg 50A, Rm 3115, Lawrence Berkeley Laboratory,
1 Cyclotron road, Berkeley, California 94720 USA.

Abstract

Hypothetical particles such as the heavy neutrino, the photino $\tilde{\gamma}$, or the sneutrino $\tilde{\nu}$ – generically called cosmions – may solve the so called missing mass problem. If they exist, the cosmions may close the Universe. In addition to their gravitational effect on cosmological scales, the cosmions may also be captured by stars and concentrate in their cores. Since cosmions are able to transport heat outside stellar cores much more efficiently than photons, they may seriously affect the thermodynamics of the inner layers of stars. We have done an exact calculation of the accretion rate of cosmions by main sequence stars and we have studied the suppression of their central convection. We concluded that central convection inside stars between 0.3 $M_\odot$ and 1 $M_\odot$ is broken in the presence of cosmions.

*On leave of absence from Laboratoire d'Annecy le Vieux de Physique des Particules, Chemin de Bellevue, BP 909, 74019 ANNECY LE VIEUX Cedex France and from Université de Chambéry, 73000, CHAMBERY, France.

1 Introduction

The dark matter puzzle is one of the most exciting issues of modern astrophysics. Solving this problem may require a close collaboration among various topics such as astrophysics and particle physics and it is mandatory that people from such different fields of research may gather and discuss as they did during this meeting.

Many theoretical and observational reasons suggest the presence of dark matter on galactic as well as extragalactic scales :

- The density of matter in the galactic disk seems to be twice the observed in the form of stars in the solar neighbourhood [1].

- The stability of spiral galaxies requires the existence of a large massive spherical halo, also needed to explain the flat rotation curves [2].

- The dynamics of galactic clusters also require a large amount of unvisible material.

- The scenarios of galaxy formation are in contradiction with observations (such as the limits on the anisotropy of the microwave background radiation) unless a large component of the mass of the universe is non baryonic [3].

The nature of dark matter is unknown, but among the many candidates for the missing mass, weak interacting massive particles are very attractive. These particles such as the heavy neutrino, the photino $\tilde{\gamma}$ or the sneutrino $\tilde{\nu}$, if they exist, have been processed during the Big-Bang in the appropriate amount to account for dark matter. These particles, generically called "cosmions", are mostly predicted by supersymmetric theories. They might be the unseen material of our galaxy and make up the galactic halo. If so, cosmions are steadily accreted by stars – our sun for instance – and concentrate in their cores. Press and Spergel [4] made an approximate calculation of the capture rate of cosmions by the sun and they showed that their concentration at the center of the star might be sufficient to solve the solar neutrino puzzle. Indeed, cosmions transport heat outside the solar core much more efficiently than photons do, leading to a reduction of the effective opacity of the central solar material and lowering the central temperature in the appropriate amount to make the solar neutrino flux compatible with the observations of Davis ($\Phi_\nu \leq 2.1$ SNU). Underground experiments are very active to try to detect cosmions either directly by their elastic scatterings upon ordinary nuclei, or indirectly through their annihilation products originating from the center of the sun.

Another interesting issue, on which we concentrate now on, is that cosmions trapped inside stars may affect their central thermal equilibrium in such a way that the inner stellar structure is changed. In order to study the thermal effects of cosmions on stars, we first must know how many of these particles can be captured by each star. That is why in section 2 we present an exact calculation [5] of the accretion rate of cosmions inside main sequence stars. Section 3 is a short overview of the consequences of energy transfer by cosmions from stellar cores. As we look for "first order" effects, we will concentrate on the suppression of central convection.

2 Capture rate of cosmions by main sequence stars

The most likely source of cosmions is the galactic halo. This last one may be roughly approximated by a nearly isothermal sphere, with a density falling as $1/r^2$ and an isotropic maxwellian distribution of velocity, with mean value $V_{halo} \simeq 300$ km/s :

$$n_x(\mathbf{r}, \mathbf{V}) = \frac{n_x(0)}{1 + \frac{r^2}{a^2}} e^{-\frac{3V^2}{2V_{halo}^2}} \tag{1}$$

where the scale length a ranges between 2 and 10 kpc. The density of cosmions in the solar neighbourghood is estimated to be around 0.01 $M_\odot/pc^3$.

The trajectories of the cosmions around some star of mass M and radius R are completely determined once the energy $E = \frac{1}{2}V^2$ and the angular momentum J with respect to the center of the star are known. Therefore, the incoming flux of cosmions (ie the number of cosmions which enter the star per second) is just given by :

$$Incoming\ Flux = 4\pi^2 n_x \left[\frac{3}{2\pi V_{halo}^2}\right]^{\frac{3}{2}} \int_{domain\ \mathcal{D}} e^{-\left(\frac{3V^2}{2V_{halo}^2}\right)} dE\ dJ^2 \tag{2}$$

The domain $\mathcal{D}$ corresponds to the trajectories the perihelion of which is of course less than the radius R of the star and is defined by :

$$E \geq 0 \quad and \quad J^2 \leq 2R^2 \cdot (E + \frac{GM}{R}) \tag{3}$$

The incoming flux gives the upper bound on the capture rate. Performing the integration of equation (2) on domain $\mathcal{D}$ leads to :

$$Incoming\ Flux\ =\ 1.04\ 10^{30}\ s^{-1} \left[\frac{m_p}{m_x}\right] \left[\frac{M}{M_\odot}\right] \left[\frac{R}{R_\odot}\right] \left[\frac{300\ km/s}{V_{halo}}\right] \left[\frac{\rho_{halo}}{0.01\ M_\odot/pc^3}\right].$$

$$\left\{ 1 + 0.1573 \left[\frac{V_{halo}}{300\ km/s}\right]^2 \left[\frac{M_\odot}{M}\right] \left[\frac{R}{R_\odot}\right] \right\} \tag{4}$$

and gives a good order of magnitude of the capture fluxes which we are going to deal with. As an example, the sun may accrete as many as 10^{30} particles each second. Equation (4) may be understood as the sum of 2 terms. The geometrical incoming flux varies as R^2 and does not depend on the mass of the star. This contribution is dominant for low density objects such as gaseous planets or red giant stars. In addition, the trajectories may be bent and focused by gravitation, leading to an extra term in the incoming flux which varies as MR. This contribution turns out to be the dominant one for main sequence stars.

Of course, not every cosmion which enters the star is captured. The particle must undergo a collision with a nucleus inside the star and lose enough energy to be unable to escape back to infinity. This results in an additional factor in the integrand of equation (2) :

$$P(E, J^2) = 1 - e^{-\int_{Trajectory} S(r)\alpha_1(r)\sigma_s n_p(r)\, dl} \tag{5}$$

$P(E,J^2)$ is just the probability of capture along the trajectory labelled by E and J^2. Obviously, $n_p(r)$ is the number density of protons at distance r from the stellar center (the main component of stars is hydrogen). σ_s is the elastic scattering cross section of cosmions upon protons. $\alpha_1(r)$ describes the thermal enhancement of the cross section due to the thermal motions of protons and is given by :

$$\alpha_1(\omega) = erf(\sqrt{\omega}) \left[1 + \frac{1}{2\omega}\right] + \frac{e^{-\omega}}{\sqrt{\pi\omega}} \tag{6}$$

where ω is just the ratio of the kinetic energies (for unit masses) of the two colliding particles :

$$\omega = \frac{m_p V^2}{2\, kT_p(r)} \tag{7}$$

Finally, S(r) is a step function which takes the value 1 if the cosmion loses enough energy in its collision with a proton to be captured, and is set equal to zero if the cosmion escapes back to infinity after having scattered.

The final step of this calculation is the description of the stellar structure. We used for this purpose the n=3 polytropic model which fits fairly well stars on the main sequence (these stars are not completely convective and a substantial part of their interior is subject to radiative transport). In addition, the strong correlation between radii and masses along the main sequence was handled by enforcing the relation :

$$\frac{R}{R_\odot} = \left(\frac{M}{M_\odot}\right)^{0.8} \tag{8}$$

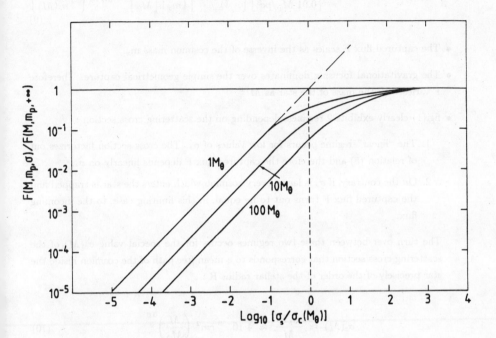

Figure 1:

The ratio of the captured flux $F(M, m_x, \sigma_s)$ to the incoming flux $F(M, m_x, \sigma_s = \infty)$ is plotted as a function of the cosmion-proton scattering cross section σ_s. $\sigma_c(M_\odot)$ is equal to $4 \cdot 10^{-36}$ cm² and corresponds to a mean free path of a cosmion inside the sun of the order of the solar radius. Three main sequence stars, with masses 1 $M_\odot$, 10 $M_\odot$ and 100 $M_\odot$ are analyzed: when the mass M of the star increases, the curve is just shifted to the right of the plot, while its shape is preserved. The cosmion mass m_x has been set equal to the proton mass m_p.

The results, displayed in fig.(1) as a function of the scattering cross section σ_s, may be summarized by the approximate relation :

$$F(M, m_x, \sigma_s) \approx 1.2\ 10^{30}\ s^{-1} \left[\frac{\rho_{halo}}{0.01\ M_\odot/pc^3}\right] \left[\frac{300\ km/s}{V_{halo}}\right] \left[\frac{m_p}{m_x}\right] \left[\frac{M}{M_\odot}\right]^{1.8} \cdot Min\left\{1, \frac{\sigma_s}{\sigma_c(M)}\right\} \quad (9)$$

- The captured flux F scales as the inverse of the cosmion mass m_x.

- The gravitational focusing dominates over the simple geometrical capture. Therefore F varies with the mass of the star as $M^{1.8}$.

- fig.(1) clearly exhibits 2 regimes, depending on the scattering cross section :

  1. The "linear" regime occurs for low values of σ_s. The cross section factorizes out of relation (5) and therefore the captured flux F depends linearly on σ_s.

  2. On the contrary, if σ_s is large, every cosmion which enters the star is trapped and the captured flux F turns out to be equal, in this limiting case, to the incoming flux.

The turn over between these two regimes occurs for the special value $\sigma_c(M)$ of the scattering cross section that corresponds to a mean free path of the cosmion inside the star precisely of the order of the stellar radius R :

$$\sigma_c(M) \approx \frac{m_p R^2}{M} \approx 4\ 10^{-36}\ cm^2 \left(\frac{M}{M_\odot}\right)^{0.6} \quad (10)$$

3   Cosmions and central convection

Cosmions are continuously captured by stars and accumulate in their cores. If cosmions annihilate, their density inside stars is of course suppressed, but it still may be possible to detect their presence through the annihilation products originating from the center of the sun. However, in this paper, we concentrate on the other issue : cosmions are stable against annihilation and their density in stellar cores steadily increases with time. As cosmions are weakly interacting particles, they transport energy more efficiently than baryons do, and therefore may compete with the other heat transport mechanisms such as radiative transport or convection. Therefore, cosmions can flatten the temperature profile at the center of stars and may induce an isothermal core. In order to understand the possible influence of cosmions on stellar structure, we are looking for the sites where the effects of these particles are potentially large. Stars the center of which are degenerate are not affected by the presence of cosmions. Indeed, electronic conduction is so large that the core of these stars is already isothermal. Therefore, we will disregard in this analysis neutron stars, white dwarfs and red giants. Stars with a radiative core are more sensitive to dark matter. The presence of cosmions leads to a decrease of their temperature gradient. This effect has been invoked to explain the solar neutrino problem. However, in this case, neither the internal structure of the star is changed, nor its lifetime. The largest effect cosmions may induce on stars is the suppression of their central convection. Not only the temperature gradient is lowered, but the structure of the core is severely changed when the star flips from convection to radiation. Moreover, convection continuously mixes matter from the outer layers with the central regions where nuclear burning takes place. As convection provides the core with fresh fuel, nuclear burning can actually take place for a much longer time than in the pure radiative case. The most important effect dark matter might have on stars is to suppress central convection – whenever it exists – and, consequently, to induce their premature death through starvation.

The rate of energy production by nuclear reactions is given by :

$$\epsilon_{nuc} = 0.562\, \rho X_H^2 \left(\frac{T}{1.5\, 10^7\, K}\right)^4 + 2.59\, \rho X_H X_{CNO} \left(\frac{T}{1.5\, 10^7\, K}\right)^{20} \quad erg.g^{-1}.s^{-1} \qquad (11)$$

where the first term refers to the PP chain and the second one describes the CNO cycle. ϵ_{nuc} is nearly constant inside the core of radius $r_{Wimps} = 0.14\, R_\odot\, \sqrt{m_p/m_x}$ where cosmions accumulate.

The nuclear energy may be radiated outside the stellar core with a luminosity L_{rad} which, at distance r from the center, is given by :

$$L_{rad} = -\frac{16\pi}{3}\, \frac{acT(r)^3}{\kappa\rho(r)}\, r^2\, \frac{dT}{dr} \qquad (12)$$

ρ is the density of matter and κ is its opacity. The product $\kappa\rho$ is just the inverse of the photon mean free path λ_γ. The convective core of a star is well described by a n = 1.5 polytrope and in this case, the radiative losses from the center may be accounted for by introducing the negative energy " production " term :

$$\epsilon_{rad} = -\frac{32\pi}{15}\, \frac{acG\mu T(0)^3}{\kappa\rho(0) R_{Boltzmann}} \quad erg.g^{-1}.s^{-1} \qquad (13)$$

where a is the energy constant of the photon gas, c is the speed of light, G is the Newton's constant of gravitation and μ is the mean molecular weight.

The cosmion heat transport takes place in two different regimes depending on the value of the Knudsen number :

$$K_n = \frac{1}{n_b \sigma_s R} = \frac{\lambda_x}{R} \qquad (14)$$

1) Large Knudsen numbers correspond to small values of the scattering cross section σ_s. Therefore, the mean free path λ_x of cosmions inside the stellar interior is much larger than the stellar radius R. Cosmions tend to have throughout the star the same mean velocity and they will be assumed here to behave as an isothermal gas with temperature T_x. T_x is just an average of the temperatures of the regions which the particles sample. The heat transferred from cosmions to baryons, due to the collisions among the two species, may also be expressed in terms of an energy production term [6] :

$$\epsilon_x = 8\sqrt{\frac{2}{\pi}} n_x n_p \frac{\sigma_s}{\rho} \frac{m_x m_p}{(m_x + m_p)^2} \sqrt{\frac{m_p k T_x + m_x kT}{m_x m_p}} \, k(T - T_x) \quad erg.g^{-1}.s^{-1} \qquad (15)$$

Requiring that cosmions are in a steady situation of thermal equilibrium with the surrounding matter :

$$\int_{star} 4\pi r^2 \epsilon_x \rho \, dr = 0 \qquad (16)$$

leads to the relation between the cosmion temperature T_x and the baryonic central temperature T_C :

$$\frac{T_x}{T_C} = 1 - \frac{3}{5} \frac{m_p}{m_x} \frac{\mu}{1\,g} \qquad (17)$$

2) Small Knudsen numbers, on the contrary, correspond to a small mean free path λ_x with respect to R. This situation is more simple to treat since cosmions are locally in thermal equilibrium with baryons (ie $T_x = T$) and the heat transport is just due to the conduction by dark matter :

$$L_x = -4\pi r^2 n_x(r) \sqrt{\frac{kT(r)}{m_x}} \frac{k}{n_p(r)\sigma_s \sqrt{\frac{3m_x + m_p}{m_x + m_p}}} \frac{dT}{dr} \qquad (18)$$

In the case of a convective core, this expression leads to :

$$\epsilon_x = \frac{8\pi}{5} \frac{G\mu}{\mathcal{N}_a} \frac{n_x(0)}{n_p(0)\sigma_s \sqrt{\frac{3m_x + m_p}{m_x + m_p}}} \sqrt{\frac{kT(0)}{m_x}} \qquad (19)$$

Our analysis simply consists in comparing the energy production or absorption terms : ϵ_{nuc} (energy produced by nuclear reactions) versus ϵ_{rad} and ϵ_x (energy absorbed by radiation and by dark matter inside a convective core). In the absence of dark matter, a convective core develops whenever radiation alone is unable to drag away all the energy generated by nuclear reactions. The star tries to radiate the central energy in increasing its temperature gradient. Whenever this last one overcomes the adiabatic gradient, which indeed corresponds to the n = 1.5 polytropic case, the center becomes unstable against convection. Relation (13) gives therefore the upper limit on the energy which may be dragged away from the stellar center through radiation losses. The condition for the existence of a convective core in the absence of dark matter is :

$$\epsilon_{nuc} \geq \epsilon_{rad} \tag{20}$$

When dark matter comes into play, it provides the stellar core with an additional cooling mechanism, and may be sufficiently large to supplement the radiation mechanism without any need for convection. This translates into :

$$\epsilon_{nuc} \leq \epsilon_{rad} + \epsilon_x \tag{21}$$

Therefore, dark matter suppresses convection in stellar cores whenever both relations (20) and (21) hold. In this case, the mixing of matter between the external layers and the very center of the star stops in the presence of cosmions, leading to a shortage of nuclear fuel inside the energy producing region. The star is starved and therefore, driven to extinction more rapidly than in the standard case.

To illustrate this effect, we deal now with main sequence stars. Recall that conditions (20) and (21) may be expressed just in terms of the central temperature T(0) and the central mass density $\rho(0)$. These quantities only depend on the mass M of the star through the somewhat phenomenological relations :

$$\begin{align}
\rho(0) &= 89.1 \ g.cm^{-3} \left(\frac{M}{M_\odot}\right)^{-1.2} \\
T(0) &= 14.6 \ 10^6 \ Kelvins \left(\frac{M}{M_\odot}\right)^{0.35} \\
\kappa &\approx 1 \ cm^2.g^{-1} \ (\ approximate \ relation \ ) \\
\mu &\simeq 0.62 \ g \ (\ for \ Pop \ I \ stars\ )
\end{align} \tag{22}$$

In addition we have set $X_H \approx 0.7$, $X_{He} \approx 0.3$ and $X_{CNO} \approx 2\%$. $X_H \approx 0.014$. The final expressions for ϵ_{nuc} and ϵ_{rad}, on the main sequence are fairly simple :

$$\epsilon_{nuc} = 22. \left(\frac{M}{M_\odot}\right)^{0.2} + 1.32 \left(\frac{M}{M_\odot}\right)^{5.8} \ erg.g^{-1}.s^{-1}$$

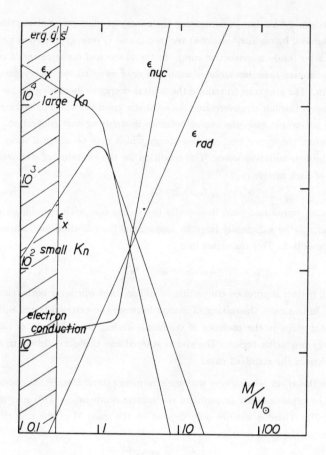

Figure 2:

The nuclear energy production rate ϵ_{nuc} and the heat absorption rates ϵ_{rad} (radiative losses) and ϵ_x (energy transported by cosmions) are plotted as a function of the mass M for main sequence stars. ϵ_x is shown in the two limiting situations of small and large Knudsen numbers. In addition, $m_x = m_p$ and $\sigma_s = 4 \cdot 10^{-36}$ cm^2. For massive stars, dark matter is unable to supplement the radiation mechanism sufficiently enough to break convection.

and
$$\epsilon_{rad} = 26.2 \left(\frac{M}{M_\odot}\right)^{2.25} \quad erg.g^{-1}.s^{-1} \qquad (23)$$

Fig.(2) is a plot of the various energy production or absorption rates as a function of the stellar mass M for main sequence stars. ϵ_{nuc} overcomes ϵ_{rad} on nearly the entire range of masses, except around 1 $M_\odot$ where radiation is sufficient to drag alone the nuclear energy produced in the core. The curve of the nuclear energy production rate exhibits two different regimes : below $\approx 1.6 M_\odot$ the PP chain is dominant but above, the CNO cycle overcomes the PP chain and ϵ_{nuc} steeply increases with M. This analysis breaks down for low mass stars – $M \leq 0.3\ M_\odot$ – where matter starts to be partly degenerate and where electronic conduction is a powerfull energy loss mechanism.

Finally, we have expressed ϵ_x in term of M. The additional feature of this calculation is that n_x depends on the capture rate F and on the lifetime τ which the star has spent on the main sequence.

$$\tau = Min\left(10^{10},\ 10^{10}\left(\frac{M}{M_\odot}\right)^{-3.6\ +\ \log\left(\frac{M}{M_\odot}\right)}\right)\ years \qquad (24)$$

This leads to the relations :

$$\epsilon_x \approx 21350\ erg.g^{-1}.s^{-1}\ Min\left(\left(\frac{M}{M_\odot}\right)^{-0.6},\ \left(\frac{M}{M_\odot}\right)^{-4.2+\log\left(\frac{M}{M_\odot}\right)}\right)\frac{\sigma_s^2}{\sigma_c^2(M_\odot)}\left[1+\frac{m_x}{m_p}\right]^{-\frac{3}{2}}$$

for large Knudsen number

and $\qquad\qquad\qquad\qquad\qquad\qquad\qquad\qquad\qquad\qquad\qquad\qquad\qquad\qquad (25)$

$$\epsilon_x \approx 6070\ erg.g^{-1}.s^{-1}\ Min\left(\left(\frac{M}{M_\odot}\right)^{0.85},\ \left(\frac{M}{M_\odot}\right)^{-2.75+\log\left(\frac{M}{M_\odot}\right)}\right)\frac{\sigma_c(M_\odot)}{\sigma_s}\sqrt{\frac{m_x+m_p}{3m_x+m_p}}$$

for small Knudsen number

In fig.(2), ϵ_x is compared with ϵ_{rad} and ϵ_{nuc}. For both low and large Knudsen numbers, dark matter is definitely unable to suppress core convection inside massive stars. The limiting effect comes from the mere fact that, despite a large capture rate F, massive stars spend too few time on the main sequence to accrete the appropriate amount of cosmions. We therefore concluded that non-annihilating cosmions may affect low mass stars ($0.3\ M_\odot \leq M \leq 0.9\ M_\odot$) by breaking their central convection, provided the scattering cross section lies in the range

[7. 10^{-38} , 7. 10^{-34}] cm². The largest thermal effect occurs for 7. 10^{-36} cm² and may be enhanced if the density of cosmions in the galactic disk is larger than the " canonical " value of 0.01 $M_\odot$ /pc³. Finally, dark matter might also affect horizontal branch stars for the same reasons, but a carefull analysis is still needed.

Aknowledgements : I would like to thank Jim Kolonko and Dave Cline for the warm and friendly atmosphere of this meeting and for the financial support they kindly provided me.

References

[1] Bahcall,J.N.,Soneira,R.M.:1980,Astrophys.J.Supp.**44**-73.

[2] Burstein,D.,Rubin,V.C.,:1985,Astrophys.J.**297**,423.
Bahcall,J.N.,Casertano,S.:1985,Astrophys.J.**293**,L7.
Blumenthal,G.,Faber,S.,Flores,R.,Primack,J.,:1986,astroph.J.**301**,27.
Flores,R.:1986, Proceedings of the Theoretical Workshop on cosmology and Particle Physics, Lawrence Berkeley laboratory, July-August 1986,Hinchliffe, I.,(ed)(World Scientific).

[3] Silk,J.: 1984, Proceedings of the Inner space-Outer space conference held at Fermilab in May 1984 (The University of Chicago Press, 1986).

[4] Press,W.H.,Spergel,D.N.,:1985,Astrophys.J.**296**,679.

[5] Bouquet,A.,Salati,P.:1987,LAPP-TH-192/87 and LPTHE 87-20, to be published in A&A.

[6] Spergel,D.N.,Press,W.H.:1985,Astrophys.J.**294**,663.

AFTER DINNER REMARKS

After Dinner Remarks

I. "Some Comments on the Problem of Funding Fundamental Research"

Dear Colleagues,

As we sit around the banquet table resting from our learned labors, I would like to tell you two fables of our time.

Our first story begins in the laboratory of Dr. Isaac Dreistein of the University. Dreistein is an expert in the recondite branch of elementary particles called thurbligs – not to be confused with jacks, aces, or quarks. Our hero is writing a proposal to the appropriate agency of the Federal Government requesting support for his research. His opus is entitled: "A Proposal in the Field of Elementary Particles – A Study of the High Spin States or Resonances of Thurbligs". After much labor and a period during which his students received a little less than their normal care and feeding, Dreistein finishes his document and sends to the agency in Washington the necessary several copies complete with budget, overhead and signatures signifying sympathetic approval of the responsible officials of his institution.

The next scene finds the proposal on the desk of a puzzled member of the Physical Sciences Branch of the Theoretical Research Division of the Elementary Particle Commission. "Ten million dollars per year for thurblig research," he cried, "Why I never even heard of such a particle." Undaunted by his ignorance of this latest important development in this rapidly advancing forefront of knowledge, our agency man proceeds in the manner usual to such cases, and gathers together a distinguished list of leading experts – the other four living thurblig experts – so that he can get their opinion of the proposed research. The people's representative – albeit a few times removed – puts three questions to his blue ribbon panel:

1) Are thurbligs important?
2) Is Dr. Dreistein's work worth supporting?

and

3) Is the University contribution adequate?

In due course, the reports of the experts are prepared and arrive at the commission. They show a remarkable unanimity on point 1), but are somewhat varied in their replies to the second question. The answer to number 3) depends on whether they contributed to or partook of the overhead charge levied against the contract. The dividing line was somewhere around the level of Dean. They expressed great enthusiasm for thurbligs. "These entities," they say, "without doubt embody the most important physical concepts since Occam's razor was used to strip the underbrush away from the face of mirror images, antiparticles, and the flow of time."

"These particles," they continued as one man, warming up to their subject, " may very well be the stuff out of which is made the glue so long sought after to explain nuclear stability."

"Are thurbligs important! Of course." say the thurblig experts, or why else would they have dedicated their lives to studying the little beasts. As I have said, the next question received a varied response.

Dr. A's report suggested that the proposal was a good one except that it failed to recognize that he had laid the groundwork in some publications several years before and, interestingly enough, he himself had mentioned the proposed line of research to Dreistein in a conversation held in the lobby of the New Yorker Hotel the preceeding January. However, said A generously, he felt that two people working on this aspect of this important field was more than warranted – although he doubted that Dreistein could use 10 million the first year, whereas other workers in the field might be better able to do so.

Dr. B's report was mailed from Shannon Airport where he was halted by fog en route to the thirty-seventh IUPAP Conference on Astrophysics and Relativity, to be held in Lake Como. He was scheduled to give a report on the interaction of thurbligs and astronomy in an invited paper entitled: "Thurblig Astronomy, What Is It?". This was a question on which he was the avowed expert since he had asked it so many times. Despite his busy schedule attending conferences and giving colloquia, B found time to comment on Dreistein's proposal. It was worth supporting he said because it might help answer the question he had been asking throughout the years. "However," he continued, "considering the field in which he was engaged, the amount requested for travel was unrealistically low."

Dr. C's review was positively ecstatic. "Dreistein," says C, a scientist and administrator at a major university, "is without doubt at least three times as creative as Einstein. Ten million was much too little for such an important field and such a good man. True, the university contribution which consisted of 1,000 dollars per year for secretarial help, was a little low but, after all, universities don't have money machines in the basement you know, and such research is inherently expensive!" There is probably no relationship between the glowing review and the fact that some years back Dreistein was a student of C's.

We omit the review of Dr. D for lack of time here tonight. Suffice it to say that it, too, was generally favorable.

This fable has several different endings which I leave to you to provide. My favorite is the one in which Dreistein gets his dough and C's advice is really taken to heart. The thurblig experts are convened in Washington and requested to make recommendations on the future course of their field. Enthusiasm mounts in the offices of the commission and it is decided as a matter of national policy that a billion dollar installation will be required for thurblig research – and a special facility is to be built for the purpose. Each state is invited to suggest sites for the new laboratory........ So, there you have it.

The listener probably shared my feeling of disquiet as this lighthearted account of serious doings was unfolded. What was the dispenser of public funds to do? After all is it reasonable to expect him to be expert in every field? What better procedure is there than to ask men who know? And how can a man be expected to be wholly objective when he is asked to support a friend or competitor's request for financial assistance? On another and equally basic point, is it in any sense reasonable to expect Drs. A through D to review the field of thurbligs in the context of all of physics and weight 10 million dollars for thurbligs against, say, an equal or even lesser amount for fluid dynamics? After all, a thurblig expert doesn't know anything about fluid dynamincs. What does the country do in the event that the total of thurblig funds requested equals the degree of freedom measured by the gross national product? Easy – we simply give less to thurbligs. But how much less? Stated in the most global terms, how do we best apportion the available resources to meet man's needs?

I would be delighted to give you the answer to these tough and increasingly pressing questions tonight but I have no ready answer. Someone has to decide how much thurbligs are worth and who is it to be? The National Academy of Sciences of the United States has taken upon itself the impossible task of writing a program for science. Congressional committees now aware of the highly visible and increasing budgetary significance of science have begun to try to understand the problem. Presumably all other activities, including the arts, will eventually write some kind of program and we'll all set forward in an integrated, planned way to a reasoned, bright, and much less uncertain future. But then what of the unforeseen discoveries which make our plans obsolete? How do we arrange to encourage them? One might counter that no plan is perfect but some plan is better than none – for without one we would have chaos in which the thurblig salesman – scientist hawks his wares in cheerful ignorance and/or disregard of the other legitimate needs of society.

But these are matters much too weighty for an evening of good food, good company and pleasant surroundings. After all, we know that there is really plenty of money for what we want to do now and it is likely that the expanding economy, fed by the very science we are discussing, will cover the expanding needs of Dreistein and others. Do we not concern ourselves unduly with non-existent peril? We all know that our field isn't so expensive and indeed even the thurblig fellow is a piker compared to the big spenders whose equipment rides rockets. Why for the cost of one rocket you could, etc., etc., etc.

Without proposing a solution to this vital and perpetual problem of the most equitable distrubution of man's resources for the greatest eventual benefit to the greatest number, I suggest that a beginning can be made in providing the broadest educational base possible so that thurblig experts do know something about fluid mechanics and even concern themselves with the interaction of the two fields. I applaud all public and professional gatherings which give some attention to achieving a broader perspective of the world around us. And finally, I suggest that we sympathetically view and be ready to help, in the continued reassessment of the role of each branch of our science as it relates to the others and even to the world at large.

* * * * * * * * * *

II. " The Sun is a Nova "

Once upon a time, there dwelled high on a mountain top a community of astrophysicists and astronomers and their families. They were a happy breed as, with their great man made eyes and heaven made brains, they searched the skies probing the mysteries of the heavens –

     Stars which shine
     Stars which do not shine
     Stars which pulse or shine steady
     Sending all manner of cosmic messengers
     To their delicate and esoteric sensors.

They were <u>indeed</u> a happy and privileged company with their eyes on the heavens.

Now it appeared one day in this scholarly and content place that one of the scientists saw strange signs in the emanations from the sun. His excitement grew apace as he recognized patterns which presaged a most remarkable event. It appeared that the sun which had nurtured and sustained life on earth for so many millenia was going to enter a nova phase that would engulf the earth as in a fiery furnace and then, mirabile dictu, the sun would return to its former familier state, the earth cleansed by fire.

With great care the scientist checked his observations and calculations in the manner he had been taught by his mentors, savoring the while the profound discovery he had made – and then he presented a report to the weekly colloquium on current research. Somehow, as it is with such things, rumors preceeded the report and the meeting place was well filled.

"What shall we do with this most remarkable information?" asked one of those assembled.

"Communicate word of the findings, so that in the time-honored manner, other astronomers on other mountains can study and check this work," said a second.

"I respectfully beg to disagree," said a third thinker. "If word of this impending temporary aberation of the sun which has with unfailing regularity heralded the arrival of each day reaches the multitude they will be most grievously disturbed."

"Yes," continued yet a fourth fellow, "let us keep these conclusions to ourselves so that we might further ponder this most unusual circumstance."

"Indeed," spoke a fifth, "knowledge is power and here we have a most singular piece of information."

"Consider," said a sixth of more philosophic bent, "the misery, the greed, the suffering that has come to infest the earth. Would it not be better to let the end come unknown and unheralded to the multitudes, for it will be sudden and brief."

"No," said a seventh, "for there must be some men on this earth with nobility of purpose and pure in heart. Should these survive the sun's temporary fluctuation then would not the world be a better place, cleansed by fire?"

"Oh," opined an eighth, "but how to make such a selection and yet preserve the word from the less deserving?"

A long silence filled the meeting place as the scholars pondered the questions and replies that had been put forth. Deep lines of concern creased the brow of a silent ninth. Then the tenth spoke with golden tongue, in measured and thoughtful words. "Is not the answer self-evident?" he intoned. "Consider those on this mountain top. Who are better selected to survive the fiery, albeit temporary, holocaust?"

They fell silent for the wisdom of the speaker was deeply compelling.

The discussion concluded, a detailed plan was rapidly formulated. The calculations showed that the sun would in a few short months enter a mild nova phase and remain in that condition for one year, after which time it would return to its former pristine and familiar regular state. Impelled by the nobility of their goal they labored diligently with a deep inner joy and contentment as they prepared an appropriate shaft and provisioned it for the period of purification.

The days flew swiftly by and finally it was prepared, and the time had come to take up residence in the shelter. They descended, one by one, until all were inside. They swung tight the heavy cover and bolted it. Secure in their retreat, they rejoiced in the knowledge and wisdom which had led them to this sacred mission and were most solicitous of each other. But this euphoria was short lived for they were men accustomed to the open sky with its vistas of twinkling stars and they found their life deep in the man-made shaft confining and irksome. As the days passed unmarked by dawn or dusk, their mental horizons shrank and they became less like men of saintly vision and more like animals. They quibbled, then quarrelled, and finally came to blows. They lost sight of the noble purpose that had taken them deep into the earth and fell to coveting the meager provisions that they had taken with them into the shaft. The quarrelling increased in intensity and became most terrible. They set upon each other; men, women, and even children, until only one man was left to contemplate the wreckage of a noble dream.

Then at long last it was time to ascend the ladder for the calculations had shown that the sun was once again in its normal state.

Slowly, with care, the lone survivor unbolted the heavy cover – and with great effort, it opened.

As the sunlight flooded the entryway, he could see fleecy clouds, he could hear birds singing – and in the distance he heard the sound of children at play.

The author wishes to state that these fables were first written over 20 years ago as Dinner Remarks for a similar occasion.